Generalized Sturmians
and
Atomic Spectra

Generalized Sturmians and Atomic Spectra

James Avery
John Avery

University of Copenhagen, Denmark

 **World Scientific**

NEW JERSEY · LONDON · SINGAPORE · BEIJING · SHANGHAI · HONG KONG · TAIPEI · CHENNAI

Published by

World Scientific Publishing Co. Pte. Ltd.
5 Toh Tuck Link, Singapore 596224
USA office: 27 Warren Street, Suite 401-402, Hackensack, NJ 07601
UK office: 57 Shelton Street, Covent Garden, London WC2H 9HE

Library of Congress Cataloging-in-Publication Data
Avery, James.
    Generalized Sturmians and atomic spectra / James Avery and John Avery.
    p. cm.
    Includes bibliographical references and index.
    ISBN-13 978-981-256-806-9 (alk. paper)
    ISBN-10 981-256-806-9 (alk. paper)
    1. Quantum theory--Mathematics. 2. Schrödinger equation. 3. Atomic spectra.
I. Title. II. Avery, John, 1933- .

QC174.17.S3 A94 2006
530.12--dc22
                                        2006048636

British Library Cataloguing-in-Publication Data
A catalogue record for this book is available from the British Library.

Copyright © 2006 by World Scientific Publishing Co. Pte. Ltd.

All rights reserved. This book, or parts thereof, may not be reproduced in any form or by any means, electronic or mechanical, including photocopying, recording or any information storage and retrieval system now known or to be invented, without written permission from the Publisher.

For photocopying of material in this volume, please pay a copying fee through the Copyright Clearance Center, Inc., 222 Rosewood Drive, Danvers, MA 01923, USA. In this case permission to photocopy is not required from the publisher.

Printed in Singapore

This book is dedicated to
Profs. Dudley R. Herschbach, Osvaldo Goscinski and Vincenzo Aquilanti

Contents

James Avery and John Avery

Preface		xi
1.	HISTORICAL BACKGROUND	1
	1.1 Sturm-Liouville theory	1
	1.2 The introduction of Sturmians into quantum theory	2
	1.3 One-electron Coulomb Sturmians	3
	1.4 Generalized Sturmians and many-particle problems	5
	1.5 Use of generalized Sturmian basis sets to solve the many-particle Schrödinger equation	6
2.	MOMENTUM SPACE AND ITERATION	9
	2.1 The d-dimensional Schrödinger equation in momentum space	9
	2.2 Momentum-space orthonormality relations for Sturmian basis sets	11
	2.3 Sturmian expansions of d-dimensional plane waves	13
	2.4 Iteration of the Schrödinger equation	14
	2.5 Generation of symmetry-adapted basis functions by iteration	16
	2.6 Solutions to the Sturmian secular equations obtained entirely by iteration	17
3.	GENERALIZED STURMIANS APPLIED TO ATOMIC SPECTRA	19
	3.1 Goscinskian configurations with weighted nuclear charges	19
	3.2 Derivation of the secular equations	24
	3.3 Symmetry-adapted basis sets for the 2-electron isoelectronic series	26

	3.4	The large-Z approximation	31
	3.5	General symmetry-adapted basis sets derived from the large-Z approximation	43
	3.6	Symmetry-adapted basis functions from iteration	44

4. AUTOIONIZING STATES — 47

	4.1	Electron correlation and the molecule-like character of autoionizing states	47
	4.2	Calculation of autoionizing states using generalized Sturmians	47
	4.3	Higher series of ^{3}S autoionizing states	51

5. CORE IONIZATION — 57

	5.1	Core ionization energies in the large-Z approximation . .	57
	5.2	Isonuclear series; piecewise-linear dependence of ΔE on N	60
	5.3	Core ionization energies for the 3-electron isoelectronic series	61

6. STRONG EXTERNAL FIELDS — 69

	6.1	External electric fields	69
	6.2	Anomalous states	70
	6.3	Polarizabilities	75
	6.4	Induced transition dipole moments	75
	6.5	External magnetic fields	76

7. RELATIVISTIC EFFECTS — 79

	7.1	Lorentz invariance and 4-vectors	79
	7.2	The Dirac equation for an electron in an external electromagnetic potential	81
	7.3	Time-independent problems	82
	7.4	The Dirac equation for an electron in the field of a nucleus	83
	7.5	Relativistic formulation of the Zeeman and Paschen-Bach effects	88
	7.6	Relativistic many-electron Sturmians	90
	7.7	A simple example	95
	7.8	Fine structure of spectral lines	101

8. MOMENTUM SPACE; THE FOCK TRANSFORMATION — 107

	8.1	One-electron Coulomb Sturmians in direct space	107
	8.2	Fourier transforms of Coulomb Sturmians	108
	8.3	The Fock projection; Hyperspherical harmonics	110
	8.4	The momentum-space orthonormality relations revisited .	111

9. HARMONIC POLYNOMIALS 117

9.1 Monomials, homogeneous polynomials, and harmonic polynomials 117
9.2 The canonical decomposition of a homogeneous polynomial 119
9.3 Generalized angular momentum 122
9.4 Hyperangular integration 124

10. HYPERSPHERICAL HARMONICS 129

10.1 The relationship between harmonic polynomials and hyperspherical harmonics 129
10.2 Construction of hyperspherical harmonics by means of harmonic projection 131
10.3 Hyperspherical harmonics in a 4-dimensional space 132
10.4 Gegenbauer polynomials 134
10.5 Hyperspherical expansion of a d-dimensional plane wave . 136
10.6 Alternative hyperspherical harmonics; The method of trees 139

11. THE MANY-CENTER PROBLEM 153

11.1 The many-center one-electron problem 153
11.2 Shibuya-Wulfman integrals 154
11.3 Shibuya-Wulfman integrals and translations 156
11.4 Matrix elements of the nuclear attraction potential 157
11.5 The Sturmian secular equations for an electron moving in a many-center potential 158
11.6 Molecular spectra 161

Appendix A INTERELECTRON REPULSION INTEGRALS 167

A.1 Procedure for evaluating the interelectron repulsion matrix 167
A.2 Separation of the integrals into radial and angular parts . 168
A.3 Evaluation of the radial integrals in terms of hypergeometric functions 169
A.4 Evaluation of the angular integrals by harmonic projection 170
A.5 Relativistic interelectron repulsion integrals 172

Appendix B GENERALIZED SLATER-CONDON RULES 175

B.1 Introduction 175
B.2 Slater determinants expressed in terms of the antisymmetrizer 175
B.3 Scalar products between configurations 176
B.4 One-electron operators 177
B.5 Two-electron operators 178

Appendix C EXPANSION OF $F(r)$ ABOUT ANOTHER CENTER 181

C.1 Expansion of a displaced function of r in terms of Gegenbauer polynomials 181
C.2 Expansion of plane waves using Gegenbauer polynomials . 182
C.3 Explicit expressions for the displaced function in terms of integrals over Bessel functions 183
C.4 An alternative method, illustrated for the case where $d=3$ 184
C.5 Closed-form differential expressions in terms of modified spherical Bessel functions 186

Appendix D THE FOCK PROJECTION 187

D.1 The one-electron Schrödinger equation in momentum space 187
D.2 The momentum-space wave equation for hydrogenlike atoms 188
D.3 Projection of momentum-space onto a 4-dimensional hypersphere 188
D.4 Expansion of the kernel in terms of hyperspherical harmonics 190

Appendix E THE GREEN'S FUNCTION OF THE SCHRÖDINGER EQUATION 193

E.1 The operator $-\Delta + p_\kappa^2$ and its Green's function 193
E.2 Conservation of symmetry under Fourier transformation . 194
E.3 Conservation of symmetry under Green's function iteration 195
E.4 Alternative representations of the Green's function 196

Appendix F CONFIGURATIONS BASED ON COULOMB STURMIAN ORBITALS 201

F.1 Coulomb Sturmian spin-orbitals 201
F.2 Generalized Shibuya-Wulfman integrals 203
F.3 An illustrative example 204
F.4 Ground states for the 2-electron isoelectronic series 205
F.5 Generalization to molecular problems 206

Appendix G NOTATION 209

Bibliography 213

Index 233

Preface

The generalized Sturmian method makes use of basis sets that are solutions to an approximate wave equation with a weighted potential. The weighting factors are chosen in such a way as to make all the members of the basis set isoenergetic. In this book we will show that when the approximate potential is taken to be that due to the attraction of the bare nucleus, the generalized Sturmian method is especially well suited for the calculation of large numbers of excited states of few-electron atoms and ions. Using the method we shall derive simple closed-form expressions that approximate the excited state energies of ions. The approximation improves with inreasing nuclear charge. The method also allows automatic generation of near-optimal symmetry adapted basis sets, and it avoids the Hartree-Fock SCF approximation.

Because of their completeness properties (discussed in Chapter 8), Sturmians have long been used as basis functions in atomic physics. Readers familiar with Sturmian basis sets will recall that they are solutions to a wave equation with a weighted potential, the weighting factors being chosen in such a way as to make the set of solutions isoenergetic. For example, Coulomb Sturmian basis sets are sets of isoenergetic square integrable solutions to

$$\left[-\frac{1}{2}\nabla^2 - \beta_n \frac{Z}{r} - E\right] \chi_{n,l,m}(\mathbf{x}) = 0$$

If the weighting factors β_n are chosen to be

$$\beta_n = \frac{kn}{Z}$$

then all of the solutions correspond to the same energy,

$$E = -\frac{k^2}{2}$$

The Coulomb Sturmians $\chi_{n,l,m}(\mathbf{x})$ are identical in form with hydrogenlike orbitals, except that Z/n is replaced everywhere with a constant, k, which is the same for all members of the basis set. They can be shown to obey a potential-weighted orthonormality relation:

$$\int d^3x \; \chi^*_{n'l'm'}(\mathbf{x}) \frac{1}{r} \chi_{nlm}(\mathbf{x}) = \frac{k}{n} \delta_{n'n} \delta_{l'l} \delta_{m'm}$$

If k is chosen in such a way that $E = -k^2/2$ is the energy of the state that is to be represented by a superposition of Sturmians, then convergence is rapid. This is because all of the members of the basis set have the correct asymptotic behavior, as will be discussed in Chapter 3. Notice that the energy E is negative. Thus Coulomb Sturmians can only be used to represent bound states. As we shall see, the same is true of generalized Sturmians.

In 1968, Osvaldo Goscinski generalized the concept of Sturmian basis sets by considering sets of square integrable solutions to N-electron wave equations of the form

$$\left[-\frac{1}{2} \sum_{j=1}^N \nabla_j^2 - \beta_\nu V_0(\mathbf{x}_1, ..., \mathbf{x}_N) - E \right] \Phi_\nu(\mathbf{x}_1, ..., \mathbf{x}_N) = 0$$

Goscinski showed that if the weighting factors β_ν are chosen in such a way as to make the set of solutions isoenergetic, then members of such a generalized Sturmian basis set obey a potential-weighted orthonormality relation analogous to that obeyed by Coulomb Sturmians. Goscinski also showed that in the case of atoms, or atomic ions, when V_0 is chosen to be the nuclear attraction potential

$$V_0(\mathbf{x}_1, ..., \mathbf{x}_N) = -\sum_{j=1}^N \frac{Z}{r_j}$$

the N-electron wave equation shown above is exactly solvable. In this book, we shall call a set of antisymmetrized isoenergetic solutions, with this choice of V_0, a set of *Goscinskian configurations*, to honor his pioneering work. In Chapters 3, 4, 5 and 6, such configurations are applied to calculations of the spectra and properties of few-electron atoms and ions. It is shown in

Chapter 3 that simple closed-form expressions can then be derived to approximate the excited state energies of large-Z members of an isoelectronic series.

The secular equations that result from the use of Goscinskian configurations have several unique features: The kinetic energy term disappears, and the nuclear attraction matrix is diagonal. Furthermore, the interelectron repulsion matrix consists of pure numbers that are energy-independent. This energy-independent interelectron repulsion matrix can be used for all states and all values of the nuclear charge Z. The roots of the secular equations are not energies, but values of a scaling parameter p_κ that is related to the spectrum of energies by

$$E_\kappa = -\frac{p_\kappa^2}{2}$$

Before solution of the secular equations, only the form of the basis functions is known, but not the scaling parameter. Solution of the secular equations results in the automatic generation of near-optimal Slater exponents appropriate for each state of the system. Thus the generalized Sturmian method offers an extremely rapid and convenient method for the calculation of large numbers of excited states of few-electron atoms and atomic ions.

When V_0 is chosen to be the nuclear attraction potential, the range of applicability of the generalized Sturmian method is limited to few-electron systems. This is because, as N increases, interelectron repulsion becomes progressively more important, and the nuclear attraction potential, by itself, resembles less and less the actual potential. Thus, as the number of electrons increases, the Goscinskian configurations (which entirely neglect interelectron repulsion) become less and less appropriate for synthesis of the actual wave function.

However, other choices of V_0 are possible. Indeed, the situation resembles that encountered in perturbation theory. One tries to represent the solution to the actual N-electron Schrödinger equation by a superposition of solutions to an approximate Schrödinger equation. The approximate equation needs to be solvable, but it should preferably be as close as possible to the true equation. It is interesting to notice that whenever V_0 has the form

$$V_0(\mathbf{x}_1, ..., \mathbf{x}_N) = \sum_{j=1}^{N} v(\mathbf{x}_j)$$

the approximate N-electron wave equation can be separated into a set of

1-electron wave equations. If spherical symmetry is assumed, the angular part of these 1-electron wave equations can be represented by spherical harmonics, and only the radial parts need special treatment. In this book, we limit the discussion to generalized Sturmian basis sets of the Goscinskian type, but it is our hope that future research will explore other choices of V_0 and thus extend the domain of applicability of the generalized Sturmian method.

Chapters 8, 9 and 10 of the book are devoted to the theory of harmonic polynomials and hyperspherical harmonics, and to their close relationship with Sturmian theory. In a famous early paper, V. Fock projected 3-dimensional momentum space onto the surface of a 4-dimensional hypersphere. Fock showed that when this projection is used, the Coulomb Sturmians correspond to 4-dimensional hyperspherical harmonics.

Vincenzo Aquilanti and his colleagues at the University of Perugia have greatly extended and deepened Fock's results, building on the strong Italian tradition of angular momentum theory founded by such important figures as Fano and Racah. The beautiful results of Aquilanti and his co-workers open up a new chapter in the theory of angular momentum and hyperangular momentum.

In the present book, we shall review and extend the hyperspherical formalisms of Fock, Aquilanti and others. We shall apply these techniques to momentum-space calculations and to evaluation of the integrals needed in relativistic calculations (Chapter 7 and Appendix A).

The wave equation in momentum space is an integral equation, and thus it can be iterated. In Appendix E, we demonstrate that symmetry is conserved under iteration of the N-electron momentum-space Schrödinger equation. This leads to a method for the automatic generation of symmetry-adapted basis sets (Chapter 2).

The final chapter, 11, deals with many-center (molecular) problems.

As the reader can imagine, we have greatly enjoyed our father-son collaboration. We felt ourselves privileged to be able to work together. It has been an enormous pleasure to write this book, and we hope that you will find equal pleasure in reading it.

Programs

The methods discussed in this book have been implemented as a shared library of routines that can be called from programs written in C, C++ and FORTRAN, as well as interfaced with Mathematica. Moreover, many of the calculations performed in this book are available as `Mathematica` notebooks that use the Generalized Sturmian Library. The programs are available at the web site

$$\text{http://sturmian.kvante.org}$$

and may be freely used as well as extended. The reader is strongly encouraged to experiment with them and to use them to perform his or her own calculations using generalized Sturmians.

Acknowledgments

We are grateful to the Universities of Perugia (Italy), Harvard (USA) and Uppsala (Sweden) for support and hospitality during research visits. We would also like to thank Professors Dudley R. Herschbach, Vincenzo Aquilanti, Osvaldo Goscinski and Sten Rettrup for many extremely enlightening conversations. Sincere thanks are also due to Dr. Andrea Caligiana and Dr. Cecilia Coletti for their important contributions.

Units

Throughout this book, atomic units are used consistently. All energies are expressed in Hartrees, and all lengths in Bohrs. The following tables provide the conversion factors to SI-units for the fundamental as well as derived units. The values are taken from [NIST, CODATA 2002].

FUNDAMENTAL ATOMIC UNITS:

Quantity	Unit	Symbol	Conversion to SI	
Energy	Hartree	E_h	$4.359744 \cdot 10^{-18}$	J
Length	Bohr	a_0	$5.2917721 \cdot 10^{-11}$	m
Mass	Electron rest mass	m_e	$9.10938 \cdot 10^{-31}$	kg
Charge	Elementary charge	e	$1.602177 \cdot 10^{-19}$	C
Angular momentum	Red. Planck const.	$\hbar$	$1.054571 \cdot 10^{-34}$	J s
Electrostatic force	Coulomb's constant	$1/(4\pi\varepsilon_0)$	$8.98755 \cdot 10^9$	$C^{-2}N\ m^2$

DERIVED ATOMIC UNITS:

Quantity	Unit	Conversion to SI	
Time	$\hbar/E_h$	$2.4188843265 \cdot 10^{-17}$	s
Force	E_h/a_0	$8.23872 \cdot 10^{-8}$	N
Velocity	$a_0 E_h/\hbar$	$2.18769126 \cdot 10^6$	m s^{-1}
Momentum	$\hbar/a_0$	$1.992852 \cdot 10^{-24}$	kg m s^{-1}
Current	$eE_h/\hbar$	$6.623618 \cdot 10^{-3}$	A
Charge density	e/a_0^3	$1.0812023 \cdot 10^{12}$	C m^{-3}
Electric field	$E_h/(ea_0)$	$5.142206 \cdot 10^{11}$	V m^{-1}
Electric dipole moment	ea_0	$8.478353 \cdot 10^{-30}$	C m
Electric polarizability	$e^2 a_0^2/E_h$	$1.6487773 \cdot 10^{-41}$	$C^2 m^2 J^{-1}$
Magnetic flux density	$\hbar/(ea_0^2)$	$2.3505173 \cdot 10^5$	T

UNIVERSAL CONSTANTS:

Quantity	Symbol	Value in A.U.	
Speed of Light	c	137.03599911	$E_h a_0/\hbar$

Chapter 1

HISTORICAL BACKGROUND

1.1 Sturm-Liouville theory

Sturmians derive their name from Sturm-Liouville theory, a branch of mathematics founded by Jacques Charles François Sturm (1803-1855) and Joseph Liouville (1809-1882). The Sturm-Liouville equation, which is named after them, can be written in the form:

$$\left[\frac{d}{dr}\left(p(r)\frac{d}{dr}\right) - q(r) + \lambda_n w(r)\right] u_n(r) = 0 \qquad (1.1)$$

Provided that boundary conditions of the form

$$\begin{aligned} u_n(a)\cos\alpha - p(a)u'(a)\sin\alpha &= 0 & 0 < \alpha < \pi \\ u_n(b)\cos\beta - p(b)u'(b)\sin\beta &= 0 & 0 < \beta < \pi \end{aligned} \qquad (1.2)$$

are imposed on a set of solutions to the Sturm-Liouville equation, and provided that $p(x)$ is differentiable, with $q(x)$ and $w(x)$ continuous, $p(x) > 0$ and $w(x) > 0$ in the interval $a < x < b$, it can be shown that

$$(\lambda_{n'}^* - \lambda_n) \int_a^b dr\ u_{n'}^*(r) w(r) u_n(r) = 0 \qquad (1.3)$$

From (1.3) it can be seen that the eigenvalues λ_n are real and that the eigenfunctions corresponding to different eigenvalues are orthogonal under an inner product weighted by $w(r)$, i.e.,

$$\int_a^b dr\ u_n^*(r) w(r) u_{n'}(r) = 0 \qquad \text{if } \lambda_{n'} \neq \lambda_n \qquad (1.4)$$

Furthermore, if the conditions mentioned above are satisfied, it can be shown that the eigenvalues are well ordered with

$$\lambda_1 < \lambda_2 < \lambda_3 < \lambda_3 < \cdots < \lambda_n < \cdots \to \infty \qquad (1.5)$$

Each eigenvalue λ_n is uniquely associated with an eigenfunction $u_n(r)$ which has exactly $n - 1$ nodes in the range $a < x < b$. The eigenfunctions $u_n(r)$ can be normalized in such a way that

$$\int_a^b dr\ u_{n'}^*(r)w(r)u_n(r) = \delta_{n',n} \qquad (1.6)$$

so that they form the orthonormal basis of a Hilbert space with weighting function $w(r)$. All second-order ordinary differential equations can be transformed into the Sturm-Liouville equation.

1.2 The introduction of Sturmians into quantum theory

One of the very early triumphs of quantum theory was the exact solution of the Schrödinger equation for hydrogenlike atoms. It was natural to try to use hydrogenlike orbitals as building blocks to represent the wave functions of more complicated atoms. However, it was soon realized that the cusps needed for accurate representation of (for example) the wave functions of heliumlike atoms and ions would require the inclusion of the continuum if the basis were to be built up of hydrogenlike orbitals. The continuum proved to be prohibitively difficult to use in practical calculations. This dilemma led Høloien, Shull and Löwdin [Shull and Löwdin, 1959] to introduce radial functions of the form

$$R_{n,l}(r) = \frac{u_{n,l}(r)}{r} \qquad (1.7)$$

where $u_{n,l}(r)$ satisfied

$$\left[\frac{d^2}{dr^2} - \frac{l(l+1)}{r^2} - \frac{k^2}{2} + \frac{kn}{r}\right] u_{n,l}(r) = 0 \qquad n = 1, 2, 3, \ldots \qquad (1.8)$$

Here k is held constant for all the members of the basis set. Shull and Löwdin were able to show that with basis sets using radial functions of this type, correct representation of the cusps could be achieved without the inclusion of the continuum. Other early authors who used functions of this type in quantum theory included Midtdal and Rotenberg [Rotenberg, 1962,1970]. Rotenberg gave the name "Sturmians" to these functions in order to call attention to their connection with Sturm-Liouville theory. The reader can verify that with the substitutions

$$p(r) \to 1$$
$$q(r) \to \frac{k^2}{2} + \frac{l(l+1)}{r^2}$$

$$\lambda_n w(r) \to \frac{kn}{r} \qquad (1.9)$$

the Sturm-Liouville equation reduces to equation (1.8).

It is interesting to notice that the Sturm-Liouville equation has sufficient flexibility to represent both the conventional type of eigenvalue equation encountered in quantum theory, where the eigenvalues λ_n are associated with the energy, and also what might be called "conjugate eigenvalue problems", where the eigenvalues λ_n are weighting factors by which the potential energy is multiplied in order to make all of the solutions correspond to a particular energy E.

1.3 One-electron Coulomb Sturmians

Rotenberg defined Sturmians as solutions of

$$\left[\frac{d^2}{dr^2} - \frac{l(l+1)}{r^2} + E - \lambda_{n,l} V_0(r) \right] u_{n,l}(r) = 0 \qquad (1.10)$$

for any $V_0(r)$ that is negative over the range $a < r < b$. For the particular set of Sturmians used by Shull and Löwdin, $E - \lambda_{n,l} V_0(r)$ has the form shown in equation (1.8), and basis sets of this type have come to be called "Coulomb Sturmians". One-electron Coulomb Sturmians have exactly the same form as the familiar hydrogenlike orbitals, except that the factor Z/n which appears in the hydrogenlike radial functions is replaced by a factor k, which is the same for all the members of the basis set. They can be written in the form

$$\chi_{nlm}(\mathbf{x}) = R_{nl}(r) Y_{lm}(\theta, \phi) \qquad (1.11)$$

where $Y_{lm}(\theta, \phi)$ is a spherical harmonic and

$$R_{nl}(r) = \mathcal{N}_{nl} (2kr)^l e^{-kr} F(l+1-n|2l+2|2kr) \qquad (1.12)$$

with

$$\mathcal{N}_{nl} = \frac{2k^{3/2}}{(2l+1)!} \sqrt{\frac{(l+n)!}{n(n-l-1)!}} \qquad (1.13)$$

In equation (1.12), $F(a|b|x)$ is a confluent hypergeometric function defined by

$$F(a|b|x) \equiv \sum_{k=0}^{\infty} \frac{\overline{a^k}}{k! \overline{b^k}} x^k = 1 + \frac{a}{b} x + \frac{a(a+1)}{2b(b+1)} x^2 + \cdots \qquad (1.14)$$

The first few Coulomb Sturmian radial functions are shown in Table 1.1 at the end of the chapter. Since the Coulomb Sturmians have the same form as the hydrogenlike orbitals with Z/n replaced by k, they must obey the hydrogenlike Schrödinger equation with the same substitution. In atomic units this becomes

$$\left[-\frac{1}{2}\nabla^2 + \frac{1}{2}k^2 - \frac{nk}{r}\right]\chi_{nlm}(\mathbf{x}) = 0 \quad (1.15)$$

For another member of the Coulomb Sturmian basis set we have

$$\left[-\frac{1}{2}\nabla^2 + \frac{1}{2}k^2 - \frac{n'k}{r}\right]\chi_{n'l'm'}(\mathbf{x}) = 0 \quad (1.16)$$

From (1.15) and (1.16) we can obtain the relationships

$$\int d^3x\, \chi^*_{n'l'm'}(\mathbf{x})\left[-\frac{1}{2}\nabla^2 + \frac{1}{2}k^2\right]\chi_{nlm}(\mathbf{x}) = \int d^3x\, \chi^*_{n'l'm'}(\mathbf{x})\frac{nk}{r}\chi_{nlm}(\mathbf{x}) \quad (1.17)$$

and

$$\int d^3x\, \chi_{nlm}(\mathbf{x})\left[-\frac{1}{2}\nabla^2 + \frac{1}{2}k^2\right]\chi^*_{n'l'm'}(\mathbf{x}) = \int d^3x\, \chi_{nlm}(\mathbf{x})\frac{n'k}{r}\chi^*_{n'l'm'}(\mathbf{x}) \quad (1.18)$$

If we subtract (1.18) from (1.17), making use of the Hermiticity of the Laplacian operator, we have

$$(n-n')\int d^3x\, \chi^*_{n'l'm'}(\mathbf{x})\frac{1}{r}\chi_{nlm}(\mathbf{x}) = 0 \quad (1.19)$$

so that

$$\int d^3x\, \chi^*_{n'l'm'}(\mathbf{x})\frac{1}{r}\chi_{nlm}(\mathbf{x}) = 0 \quad \text{if } n \neq n' \quad (1.20)$$

Finally, using the fact that the hydrogenlike orbitals obey the Virial theorem, and also making use of the orthonormality of the spherical harmonics we obtain

$$\int d^3x\, \chi^*_{n'l'm'}(\mathbf{x})\frac{1}{r}\chi_{nlm}(\mathbf{x}) = \frac{k}{n}\delta_{n'n}\delta_{l'l}\delta_{m'm} \quad (1.21)$$

which is the potential-weighted orthonormality relation obeyed by Coulomb Sturmian basis sets.

1.4 Generalized Sturmians and many-particle problems

In a pioneering 1968 paper, Osvaldo Goscinski [Goscinski, 1968, 2003] generalized the concept of Sturmians. He regarded Sturmians as isoenergetic solutions to a general d-dimensional Schrödinger-like equation of the form

$$\left[-\frac{1}{2}\Delta + \beta_\nu V_0(\mathbf{x}) - E_\kappa\right] \Phi_\nu(\mathbf{x}) = 0 \qquad (1.22)$$

with a weighted potential $\beta_\nu V_0(\mathbf{x})$, the weighting constants β_ν being chosen in such a way as to make all of the solutions correspond to the same energy, E_κ. Equation (1.22) could be applied to N particles of different masses m_j with

$$\Delta \equiv \sum_{j=1}^{N} \frac{1}{m_j} \nabla_j^2 \qquad (1.23)$$

For a collection of N electrons, one can let

$$\Delta \equiv \sum_{j=1}^{d} \frac{\partial^2}{\partial x_j^2} \qquad (1.24)$$

where $d = 3N$ and

$$\mathbf{x} = (x_1, x_2, ..., x_d) \qquad (1.25)$$

since, in atomic units, $m_j = 1$ for all the electrons. Goscinski was able to show (by an argument similar to equations (1.15)-(1.21)), that generalized Sturmians obey a potential-weighted orthogonality relation of the form

$$\int d\mathbf{x}\ \Phi_{\nu'}^*(\mathbf{x}) V_0(\mathbf{x}) \Phi_\nu(\mathbf{x}) = 0 \qquad \text{if } \beta_{\nu'} \neq \beta_\nu \qquad (1.26)$$

We should notice that in equations (1.22) and (1.26), ν stands for a set of quantum numbers, some of which influence the value of β_ν, while others do not. We can call these respectively *major* and *minor* quantum numbers. Orthogonality with respect to the minor quantum numbers can often be established by means of symmetry properties. When this is not possible, the members of a Sturmian basis set can be made orthogonal with respect to the remaining minor quantum numbers by means of for example Graham-Schmidt or Löwdin orthogonalization. We shall normalize our generalized Sturmian basis sets in such a way that the potential-weighted orthonormality relations take the form

$$\int d\mathbf{x}\ \Phi_{\nu'}^*(\mathbf{x}) V_0(\mathbf{x}) \Phi_\nu(\mathbf{x}) = -\delta_{\nu',\nu} \frac{p_\kappa^2}{\beta_\nu} \qquad (1.27)$$

where
$$p_\kappa^2 \equiv -2E_\kappa \qquad (1.28)$$

This type of normalization is convenient because

(1) It makes (1.27) reduce to (1.21) for the case of Coulomb Sturmians (with $d=3$, $p_\kappa = k$ and $\beta_\nu = nk$).
(2) It makes the generalized Sturmians fit more naturally into the theory of Sobolev spaces.
(3) The special Goscinskian configurations, which will be introduced in Chapter 3, are already properly normalized.

1.5 Use of generalized Sturmian basis sets to solve the many-particle Schrödinger equation

The non-relativistic many-particle Schrödinger equation can be written in the form
$$\left[-\frac{1}{2}\Delta + V(\mathbf{x}) - E_\kappa \right] \Psi_\kappa(\mathbf{x}) = 0 \qquad (1.29)$$

where atomic units are used and where Δ is defined by (1.23) or (1.24). We now expand the wave function $\Psi_\kappa(\mathbf{x})$ in terms of a generalized Sturmian basis set.
$$\Psi_\kappa(\mathbf{x}) = \sum_\nu \Phi_\nu(\mathbf{x}) B_{\nu,\kappa} \qquad (1.30)$$

Substituting this expansion into the Schrödinger equation, and using the fact that all the members of our basis set obey equation (1.22), we obtain
$$\sum_\nu \left[-\frac{1}{2}\Delta + V(\mathbf{x}) - E_\kappa \right] \Phi_\nu(\mathbf{x}) B_{\nu,\kappa}$$
$$= \sum_\nu \left[V(\mathbf{x}) - \beta_\nu V_0(\mathbf{x}) \right] \Phi_\nu(\mathbf{x}) B_{\nu,\kappa} = 0 \qquad (1.31)$$

If we multiply (1.31) on the left by a conjugate function from our basis set and integrate over all the coordinates, we obtain
$$\sum_\nu \int d\mathbf{x}\, \Phi_{\nu'}^*(\mathbf{x}) \left[V(\mathbf{x}) - \beta_\nu V_0(\mathbf{x}) \right] \Phi_\nu(\mathbf{x}) B_{\nu,\kappa} = 0 \qquad (1.32)$$

We next introduce the notation

$$T_{\nu',\nu} \equiv -\frac{1}{p_\kappa} \int dx \ \Phi^*_{\nu'}(\mathbf{x}) V(\mathbf{x}) \Phi_\nu(\mathbf{x}) \tag{1.33}$$

Finally, making use of the potential-weighted orthonormality relations (1.27), we obtain the Sturmian secular equations.

$$\sum_\nu \left[T_{\nu',\nu} - p_\kappa \delta_{\nu',\nu} \right] B_{\nu,\kappa} = 0 \tag{1.34}$$

These secular equations have several remarkable features that will be discussed in more detail in later chapters. For the moment, it is interesting to notice that the kinetic energy term has disappeared. Furthermore, the eigenvalues are not energies, but values of the "scaling parameter" p_κ, which is related to the energy spectrum through equation (1.28).

Table 1.1: One-electron Coulomb Sturmian radial functions. If k is replaced by Z/n they are identical to the familiar hydrogenlike radial wave functions.

n	l	$R_{n,l}(r)$
1	0	$2k^{3/2}e^{-kr}$
2	0	$2k^{3/2}(1-kr)e^{-kr}$
2	1	$\dfrac{2k^{3/2}}{\sqrt{3}}\, kr\, e^{-kr}$
3	0	$2k^{3/2}\left(1 - 2kr + \dfrac{2(kr)^2}{3}\right)e^{-kr}$
3	1	$2k^{3/2}\dfrac{2\sqrt{2}}{3}\, kr\left(1 - \dfrac{kr}{2}\right)e^{-kr}$
3	2	$2k^{3/2}\dfrac{\sqrt{2}}{3\sqrt{5}}\,(kr)^2\, e^{-kr}$

Chapter 2

MOMENTUM SPACE AND ITERATION

2.1 The d-dimensional Schrödinger equation in momentum space

In Chapter 1, we discussed the non-relativistic Schrödinger equation for N electrons moving in a potential V. This was expressed as a d-dimensional wave equation, where $d = 3N$. We would now like to rewrite the wave equation as an integral equation in momentum space. To do this, we let

$$e^{i\mathbf{p}\cdot\mathbf{x}} \equiv e^{i(p_1 x_1 + ... + p_d x_d)} \tag{2.1}$$

be a d-dimensional plane wave. Then the N-electron wave function $\Psi_\kappa(\mathbf{x})$ and its Fourier transform $\Psi_\kappa^t(\mathbf{p})$ are related by

$$\Psi_\kappa(\mathbf{x}) = \frac{1}{(2\pi)^{d/2}} \int dp\ e^{i\mathbf{p}\cdot\mathbf{x}} \Psi_\kappa^t(\mathbf{p})$$
$$\Psi_\kappa^t(\mathbf{p}) = \frac{1}{(2\pi)^{d/2}} \int dx\ e^{-i\mathbf{p}\cdot\mathbf{x}} \Psi_\kappa(\mathbf{x}) \tag{2.2}$$

where

$$\Psi_\kappa(\mathbf{x}) \equiv \Psi_\kappa(x_1, ..., x_d)$$
$$\Psi_\kappa^t(\mathbf{p}) \equiv \Psi_\kappa^t(p_1, ..., p_d)$$
$$dx \equiv dx_1 dx_2 ... dx_d$$
$$dp \equiv dp_1 dp_2 ... dp_d \tag{2.3}$$

If we let

$$p_\kappa^2 \equiv -2E_\kappa \tag{2.4}$$

and

$$\Delta \equiv \sum_{j=1}^{d} \frac{\partial^2}{\partial x_j^2} \tag{2.5}$$

then the N-electron Schrödinger equation can be written in the form

$$\left[-\Delta + p_\kappa^2 + 2V(\mathbf{x})\right] \Psi_\kappa(\mathbf{x}) = 0 \tag{2.6}$$

Substituting the expression for the wave function in terms of its Fourier transform (2.2) into (2.6), we have

$$\int dp\, e^{i\mathbf{p}\cdot\mathbf{x}} \left[p^2 + p_\kappa^2 + 2V(\mathbf{x})\right] \Psi_\kappa^t(\mathbf{p}) = 0 \tag{2.7}$$

since $-\Delta$ acting on the plane wave brings down the factor p^2. If we now multiply (2.7) by $e^{-i\mathbf{p}'\cdot\mathbf{x}}$ and integrate over the space coordinates, we obtain

$$\int dp\, (p^2 + p_\kappa^2) \Psi_\kappa^t(\mathbf{p}) \int dx\, e^{i(\mathbf{p}-\mathbf{p}')\cdot\mathbf{x}}$$
$$+ 2\int dp \int dx\, e^{i(\mathbf{p}-\mathbf{p}')\cdot\mathbf{x}} V(\mathbf{x}) \Psi_\kappa^t(\mathbf{p}) = 0 \tag{2.8}$$

Then, remembering that

$$\int dx\, e^{i(\mathbf{p}-\mathbf{p}')\cdot\mathbf{x}} = (2\pi)^d \delta(\mathbf{p}-\mathbf{p}') \tag{2.9}$$

we have

$$(2\pi)^d \int dp\, \delta(\mathbf{p}-\mathbf{p}')(p^2 + p_\kappa^2) \Psi_\kappa^t(\mathbf{p})$$
$$+ 2\int dp \int dx\, e^{i(\mathbf{p}-\mathbf{p}')\cdot\mathbf{x}} V(\mathbf{x}) \Psi_\kappa^t(\mathbf{p}) = 0 \tag{2.10}$$

Using the d-dimensional Dirac delta function to perform the first p-integration in (2.10), we obtain the momentum-space Schrödinger equation:

$$(p'^2 + p_\kappa^2)\Psi_\kappa^t(\mathbf{p}') = -\frac{2}{(2\pi)^{d/2}} \int dp\, V^t(\mathbf{p}'-\mathbf{p}) \Psi_\kappa^t(\mathbf{p}) \tag{2.11}$$

where

$$V^t(\mathbf{p}'-\mathbf{p}) \equiv \frac{1}{(2\pi)^{d/2}} \int dx\, e^{-i(\mathbf{p}'-\mathbf{p})\cdot\mathbf{x}} V(\mathbf{x}) \tag{2.12}$$

2.2 Momentum-space orthonormality relations for Sturmian basis sets

We now introduce a set of generalized Sturmian basis functions. As was discussed in Chapter 1, such a set is defined to be a set of square integrable solutions to the Schrödinger equation with a weighted zeroth-order potential, $\beta_\nu V_0(\mathbf{x})$, the weighting factor β_ν being chosen in such a way as to make all of the solutions correspond to the same energy E_κ. In other words, the generalized Sturmian basis set is a set of solutions to the d-dimensional wave equation

$$\left[-\frac{1}{2}\Delta + \beta_\nu V_0(\mathbf{x}) - E_\kappa\right] \Phi_\nu(\mathbf{x}) = 0 \quad (2.13)$$

where

$$-\frac{1}{2}\Delta \equiv -\frac{1}{2}\sum_{j=1}^N \frac{1}{m_j}\nabla_j^2 \quad (2.14)$$

and where the weighting factors β_ν are especially chosen so that all the functions in the basis set will be isoenergetic. In Chapter 1, we showed that such a set of generalized Sturmian basis functions obey a potential-weighted orthonormality relation in direct space

$$\int d\mathbf{x} \; \Phi^*_{\nu'}(\mathbf{x})V_0(\mathbf{x})\Phi_\nu(\mathbf{x}) = \delta_{\nu',\nu}\frac{2E_\kappa}{\beta_\nu} = -\delta_{\nu',\nu}\frac{p_\kappa^2}{\beta_\nu} \quad (2.15)$$

where

$$p_\kappa^2 \equiv -2E_\kappa \quad (2.16)$$

We would now like to find the momentum-space orthonormality relations obeyed by Fourier transforms of the generalized Sturmian basis set. Because the Fourier Transform is unitary, the inner product of any two functions in L_2 is preserved under the operation of taking their Fourier transforms, i.e.

$$\int d\mathbf{x} \; f^*(\mathbf{x})g(\mathbf{x}) = \int d\mathbf{p} \; f^{t*}(\mathbf{p})g^t(\mathbf{p}) \quad (2.17)$$

Using this well-known relationship with $f^*(\mathbf{x}) = \Phi^*_{\nu'}(\mathbf{x})$ and $g(\mathbf{x}) = V_0(\mathbf{x})\Phi_\nu(\mathbf{x})$, we have

$$\int d\mathbf{x} \; \Phi^*_{\nu'}(\mathbf{x})V_0(\mathbf{x})\Phi_\nu(\mathbf{x}) = \int d\mathbf{p} \; \Phi^{t*}_{\nu'}(\mathbf{p})\left[V_0\Phi_\nu\right]^t(\mathbf{p}) \quad (2.18)$$

In order to evaluate $[V_0 \Phi_\nu]^t (\mathbf{p})$, we remember the Fourier convolution theorem, which states that the Fourier transform of the product of two functions is the convolution of their Fourier transforms. Thus if a and b are any two functions in L_2,

$$[ab]^t (\mathbf{p}) \equiv \frac{1}{(2\pi)^{d/2}} \int d\mathbf{x} \; e^{-i\mathbf{p}' \cdot \mathbf{x}} a(\mathbf{x}) b(\mathbf{x}) = \frac{1}{(2\pi)^{d/2}} \int d\mathbf{p} \; a^t(\mathbf{p}' - \mathbf{p}) b^t(\mathbf{p}') \tag{2.19}$$

Letting $a(\mathbf{x}) = V_0(\mathbf{x})$ and $b(\mathbf{x}) = \Phi_\nu(\mathbf{x})$ we have

$$[V_0 \Phi_\nu]^t (\mathbf{p}') = \frac{1}{(2\pi)^{d/2}} \int d\mathbf{p} \; V_0^t(\mathbf{p}' - \mathbf{p}) \Phi_\nu^t(\mathbf{p}) \tag{2.20}$$

Since the momentum-space integral equation corresponding to (2.13) has the form

$$(p'^2 + p_\kappa^2) \Phi_\nu^t(\mathbf{p}') = -\frac{2\beta_\nu}{(2\pi)^{d/2}} \int d\mathbf{p} \; V_0^t(\mathbf{p}' - \mathbf{p}) \Phi_\nu^t(\mathbf{p}) \tag{2.21}$$

it follows that

$$[V_0 \Phi_\nu]^t (\mathbf{p}) = -\frac{(p^2 + p_\kappa^2)}{2\beta_\nu} \Phi_\nu^t(\mathbf{p}) \tag{2.22}$$

Finally, substituting (2.22) into (2.18), we obtain the momentum-space orthonormality relations for a set of generalized Sturmian basis functions:

$$\int d\mathbf{p} \; \Phi_{\nu'}^{t*}(\mathbf{p}) \left(\frac{p^2 + p_\kappa^2}{2p_\kappa^2} \right) \Phi_\nu^t(\mathbf{p}) = \delta_{\nu',\nu} \tag{2.23}$$

Because all of the functions $\Phi_\nu(\mathbf{x})$ in the generalized Sturmian basis set obey equation (2.13), the potential-weighted direct space orthonormality relations shown in equation (2.15) can be rewritten in the form

$$\int d\mathbf{x} \; \Phi_{\nu'}^*(\mathbf{x}) \left(\frac{-\Delta + p_\kappa^2}{2p_\kappa^2} \right) \Phi_\nu(\mathbf{x}) = \delta_{\nu',\nu} \tag{2.24}$$

so that the momentum-space and direct-space orthonormality relations can be seen to be related to each other in a symmetrical way. These weighted orthonormality relations in $L_2(\mathbb{R}^d)$ are the usual orthonormality relations in the Sobolev space $W_2^{(1)}(\mathbb{R}^d)$ (see [Weniger, 1985]).

2.3 Sturmian expansions of d-dimensional plane waves

If the set of generalized Sturmian basis functions is complete in the sense of spanning the Sobolev space $W_2^{(1)}(\mathbb{R}^d)$, we can use it to construct a weakly convergent expansion of a d-dimensional plane wave (valid only in the sense of distributions). Suppose that we let

$$e^{i\mathbf{p}\cdot\mathbf{x}} = \left(\frac{p_\kappa^2 + p^2}{2p_\kappa^2}\right) \sum_\nu \Phi_\nu^{t*}(\mathbf{p}) a_\nu \qquad (2.25)$$

We can then determine the coefficients a_ν by means of the orthonormality relations (2.23). Multiplying (2.25) on the left by $\Phi_{\nu'}^{t*}(\mathbf{p})$ and integrating over dp making use of (2.23), we obtain

$$\int dp\, e^{i\mathbf{p}\cdot\mathbf{x}} \Phi_{\nu'}^{t*}(\mathbf{p}) = \sum_\nu \delta_{\nu',\nu} a_\nu = a_{\nu'} \qquad (2.26)$$

so that

$$a_\nu = \int dp\, e^{i\mathbf{p}\cdot\mathbf{x}} \Phi_\nu^t(\mathbf{p}) = (2\pi)^{d/2} \Phi_\nu(\mathbf{x}) \qquad (2.27)$$

Thus finally we obtain an expansion of the form

$$e^{i\mathbf{p}\cdot\mathbf{x}} = (2\pi)^{d/2} \left(\frac{p_\kappa^2 + p^2}{2p_\kappa^2}\right) \sum_\nu \Phi_\nu^{t*}(\mathbf{p}) \Phi_\nu(\mathbf{x}) \qquad (2.28)$$

If the set of generalized Sturmians $\Phi_\nu(\mathbf{x})$ does not span $W_2^{(1)}(\mathbb{R}^d)$, equation (2.28) becomes

$$P\left[e^{i\mathbf{p}\cdot\mathbf{x}}\right] = (2\pi)^{d/2} \left(\frac{p_\kappa^2 + p^2}{2p_\kappa^2}\right) \sum_\nu \Phi_\nu^{t*}(\mathbf{p}) \Phi_\nu(\mathbf{x}) \qquad (2.29)$$

where $P\left[e^{i\mathbf{p}\cdot\mathbf{x}}\right]$ is the projection of the d-dimensional plane wave onto the subspace spanned by the set $\{\Phi_\nu(\mathbf{x})\}$. For example, if we are considering a system of N electrons, with $d = 3N$, the generalized Sturmian basis set might be antisymmetric with respect to exchange of the N electron coordinates but otherwise complete. In that case, $P\left[e^{i\mathbf{p}\cdot\mathbf{x}}\right]$ would represent the projection of the plane wave onto that part of Hilbert space corresponding to functions of $\mathbf{x}$ that are antisymmetric with respect to exchange of the N electron coordinates. Neither the expansion shown in equation (2.28) nor that shown in equation (2.29) is point-wise convergent. In other words, we cannot perform the sums shown on the right-hand sides of these equations and expect them to give point-wise convergent representations of the plane wave or its projection. However, the expansions are valid in the sense of distributions. Thus, for example, if

$$f(\mathbf{p}) = \sum_{\nu'} \Phi^t_{\nu'}(\mathbf{p}) c_{\nu'} \qquad (2.30)$$

then

$$\frac{1}{(2\pi)^{d/2}} \int dp \, e^{i\mathbf{p}\cdot\mathbf{x}} f(\mathbf{p}) = \frac{1}{(2\pi)^{d/2}} \int dp \, e^{i\mathbf{p}\cdot\mathbf{x}} \sum_{\nu'} \Phi^t_{\nu'}(\mathbf{p}) c_{\nu'} \qquad (2.31)$$

Replacing the plane wave in (2.31) by the expression given in (2.28), and making use of the reciprocal-space orthonormality relations (2.23), we obtain:

$$\frac{1}{(2\pi)^{d/2}} \int dp \, e^{i\mathbf{p}\cdot\mathbf{x}} f(\mathbf{p}) = \sum_{\nu',\nu} c_{\nu'} \Phi_\nu(\mathbf{x}) \int dp \, \Phi^{t*}_\nu(\mathbf{p}) \left(\frac{p_\kappa^2 + p^2}{2 p_\kappa^2} \right) \Phi^t_{\nu'}(\mathbf{p})$$

$$= \sum_{\nu',\nu} \Phi_\nu(\mathbf{x}) c_{\nu'} \delta_{\nu',\nu} = \sum_\nu \Phi_\nu(\mathbf{x}) c_\nu \qquad (2.32)$$

This result can be seen to be valid if the Fourier transforms of both sides are compared with equation (2.30).

2.4 Iteration of the Schrödinger equation

We are now in a position to iterate various integral forms of the Schrödinger equation. Using the Fourier convolution theorem, (2.19), we can rewrite the momentum-space wave equation (2.11) in the form

$$(p_\kappa^2 + p^2) \Psi^t_\kappa(\mathbf{p}) = -\frac{2}{(2\pi)^{d/2}} \int dx \, e^{-i\mathbf{p}\cdot\mathbf{x}} V(\mathbf{x}) \Psi_\kappa(\mathbf{x}) \qquad (2.33)$$

Dividing both sides of (2.33) by $(p_\kappa^2 + p^2)$, multiplying by $e^{i\mathbf{p}\cdot\mathbf{x}'}$ and integrating over dp, we have

$$\int dp \, e^{i\mathbf{p}\cdot\mathbf{x}'} \Psi^t_\kappa(\mathbf{p}) = -\frac{2}{(2\pi)^{d/2}} \int dx \int dp \, \frac{e^{i\mathbf{p}\cdot(\mathbf{x}'-\mathbf{x})}}{(p_\kappa^2 + p^2)} V(\mathbf{x}) \Psi_\kappa(\mathbf{x}) \qquad (2.34)$$

This can be rewritten in the form

$$\Psi_\kappa(\mathbf{x}') = -2 \int dx \, G(\mathbf{x}' - \mathbf{x}) V(\mathbf{x}) \Psi_\kappa(\mathbf{x}) \qquad (2.35)$$

where

$$G(\mathbf{x}' - \mathbf{x}) = \frac{1}{(2\pi)^d} \int dp \, \frac{e^{i\mathbf{p}\cdot(\mathbf{x}'-\mathbf{x})}}{p_\kappa^2 + p^2} \qquad (2.36)$$

An alternative derivation of equations (2.35) and (2.36) is given in Appendix E.

If the set of generalized Sturmian basis functions $\Phi_\nu(\mathbf{x})$ is complete, we can substitute the complex conjugate of (2.28) into (2.33). This gives us

$$\Psi_\kappa^t(\mathbf{p}) = -\frac{1}{p_\kappa^2} \sum_\nu \Phi_\nu^t(\mathbf{p}) \int dx \, \Phi_\nu^*(\mathbf{x}) V(\mathbf{x}) \Psi_\kappa(\mathbf{x}) \qquad (2.37)$$

Taking the Fourier transform on both sides of (2.37), we have

$$\Psi_\kappa(\mathbf{x}') = -\frac{1}{p_\kappa^2} \sum_\nu \Phi_\nu(\mathbf{x}') \int dx \, \Phi_\nu^*(\mathbf{x}) V(\mathbf{x}) \Psi_\kappa(\mathbf{x}) \qquad (2.38)$$

Finally, if we compare (2.38) with (2.35), we can make the identification (in the sense of distributions)

$$G(\mathbf{x}' - \mathbf{x}) = \frac{1}{2p_\kappa^2} \sum_\nu \Phi_\nu(\mathbf{x}') \Phi_\nu^*(\mathbf{x}) \qquad (2.39)$$

In other words, for any complete set of generalized Sturmian basis functions, we can make the identification

$$\frac{1}{2p_\kappa^2} \sum_\nu \Phi_\nu(\mathbf{x}') \Phi_\nu^*(\mathbf{x}) = \frac{1}{(2\pi)^d} \int dp \, \frac{e^{i\mathbf{p}\cdot(\mathbf{x}'-\mathbf{x})}}{p_\kappa^2 + p^2} \qquad (2.40)$$

which is valid in the sense of distributions.

Suppose that we now wish to iterate the Schrödinger equation in the form shown in equation (2.38). Substituting an initial solution into the integral on the right-hand side, we obtain a first-iterated solution. This can in turn be substituted into the integral on the right-hand side, yielding a second-iterated solution, and so on.

$$\Psi_\kappa^{(1)}(\mathbf{x}') = -\frac{1}{p_\kappa^2} \sum_\nu \Phi_\nu(\mathbf{x}') \int dx \, \Phi_\nu^*(\mathbf{x}) V(\mathbf{x}) \Psi_\kappa^{(0)}(\mathbf{x})$$

$$\Psi_\kappa^{(2)}(\mathbf{x}') = -\frac{1}{p_\kappa^2} \sum_\nu \Phi_\nu(\mathbf{x}') \int dx \, \Phi_\nu^*(\mathbf{x}) V(\mathbf{x}) \Psi_\kappa^{(1)}(\mathbf{x})$$

$$\vdots \quad \vdots \quad \vdots \qquad (2.41)$$

If we now let

$$T_{\nu',\nu} \equiv -\frac{1}{p_\kappa} \int dx\, \Phi^*_{\nu'}(\mathbf{x}) V(\mathbf{x}) \Phi_\nu(\mathbf{x}) \qquad (2.42)$$

and

$$\Psi^{(i)}_\kappa(\mathbf{x}) = \sum_{\nu \in I_i} \Phi_\nu(\mathbf{x}) B^{(i)}_{\nu,\kappa} \qquad (2.43)$$

it follows that

$$B^{(i+1)}_{\nu',\kappa} = \frac{1}{p_\kappa} \sum_{\nu \in I_i} T_{\nu',\nu} B^{(i)}_{\nu,\kappa} \qquad \nu' \in I_{i+1} \qquad (2.44)$$

Thus the iteration can easily be performed in practice provided that we are able to evaluate the matrix elements $T_{\nu',\nu}$.

2.5 Generation of symmetry-adapted basis functions by iteration

An initial solution can be obtained by solving the Sturmian secular equations

$$\sum_{\nu \in I_0} [T_{\nu',\nu} - p_\kappa \delta_{\nu',\nu}] B^{(0)}_{\nu,\kappa} = 0 \qquad (2.45)$$

with a truncated basis set contained in the domain I_0. Equation (2.44) can then be used as a criterion for automatic selection of a larger basis set to be used in a more accurate version of the secular equations. If the initial solution has a symmetry corresponding to one of the irreducible representations of the symmetry group of $V(\mathbf{x})$, the first-iterated solution will be of the same symmetry, as is discussed in Appendix E, provided that the domain I_1 is appropriately chosen. Thus iteration can be used to construct symmetry-adapted basis sets. The appropriate domains I_0 and I_1 for generating symmetry-adapted basis sets of the Russell-Saunders type will be discussed in Chapter 3.

If we substitute (2.29) into (2.33), we can see that equations (2.43)-(2.45) can be used to iterate the N-electron Schrödinger equation even when the generalized Sturmian basis set $\Phi_\nu(\mathbf{x})$ is not complete, but of course the iterated solutions will never leave the part of Sobolev space spanned by the basis set.

2.6 Solutions to the Sturmian secular equations obtained entirely by iteration

It is interesting to ask whether solutions to the Sturmian secular equations (1.34) could in principle be obtained entirely by iteration, without any diagonalization step. To answer this question, let us suppose that our initial trial function is a linear superposition of the true solutions, so that

$$B_\nu^{(0)} = \sum_\kappa c_\kappa B_{\nu,\kappa} \qquad (2.46)$$

Then

$$B_{\nu'}^{(i)} \sim \sum_\nu T^i_{\nu',\nu} B_\nu^{(0)} = \sum_\kappa c_\kappa \sum_\nu T^i_{\nu',\nu} B_{\nu,\kappa} \qquad (2.47)$$

where the superscript i on the T matrix means that it has been raised to the ith power. With the help of the secular equations (1.34) we find that

$$B_{\nu'}^{(i)} \sim \sum_\kappa c_\kappa (p_\kappa)^i B_{\nu,\kappa} \qquad (2.48)$$

As the number of iterations i becomes very large, the eigenfunction with the largest value of $|p_\kappa|$ in the original trial function dominates in the iterated solution. This does not necessarily mean that a multiply iterated solution approaches the ground state of a system. For example, if the initial trial function has a symmetry that is different from the symmetry of the ground state, then the multiply iterated solution will retain that symmetry and will converge to the solution contained in the trial function of equation (2.46) whose value of $|p_\kappa|$ is the largest. The convergence of basis-set-free Green's function iterations (Appendix E) will be similar. Efficient methods for obtaining iterated solutions for secular equations of large dimension have been developed by Cornelius Lanczos [Lanczoz, 1950], Isaiah Shavitt, Ernest Davidson and others.

Chapter 3

GENERALIZED STURMIANS APPLIED TO ATOMIC SPECTRA

3.1 Goscinskian configurations with weighted nuclear charges

In atomic units, the non-relativistic, time-independent Schrödinger equation for N particles moving in a potential V can be written in the form

$$\left[-\frac{1}{2}\Delta + V(\mathbf{x}) - E_\kappa\right] \Psi_\kappa(\mathbf{x}) = 0 \qquad (3.1)$$

where $\mathbf{x} \in \mathbb{R}^{3N}$ represents the N Cartesian coordinates $\mathbf{x}_i \in \mathbb{R}^3$ of the particles, and where the operator

$$-\frac{1}{2}\Delta \equiv -\frac{1}{2} \sum_{j=1}^{d} \frac{\partial^2}{\partial x_j^2} \qquad (3.2)$$

is the kinetic energy operator. It is customary to build the solutions to equation (3.1) as superpositions of some known set of basis functions

$$\Psi_\kappa(\mathbf{x}) = \sum_\nu \Phi_\nu(\mathbf{x}) B_{\nu,\kappa} \qquad (3.3)$$

In the generalized Sturmian method [Avery, 2000], [Aquilanti and Avery, 2001], [Goscinski, 2003], the basis functions Φ_ν are solutions to an approximate Schrödinger equation of the form

$$\left[-\frac{1}{2}\Delta + \beta_\nu V_0(\mathbf{x}) - E_\kappa\right] \Phi_\nu(\mathbf{x}) = 0 \qquad (3.4)$$

where V_0 is a potential chosen in such a way that equation (3.4) can be solved to a good approximation. The constant weighting factors β_ν are constructed such that all of the functions Φ_ν are isoenergetic, corresponding to the energy E_κ of the state that they are used to represent. It is important that V_0 resemble the actual potential V to the greatest extent possible, and especially that $V_0(\mathbf{x}) \to 0$ when $V(\mathbf{x}) \to 0$. In this case, equations (3.1) and (3.4) obey the same N-particle Schrödinger equation in the limit where both $V(\mathbf{x})$ and $V_0(\mathbf{x})$ go to zero:

$$\left[-\tfrac{1}{2}\Delta + \beta_\nu V_0(\mathbf{x}) - E_\kappa\right]\Phi_\nu(\mathbf{x}) \to \left[-\tfrac{1}{2}\Delta - E_\kappa\right]\Phi_\nu(\mathbf{x}) \text{ for } V_0(\mathbf{x}) \to 0$$

$$\left[-\tfrac{1}{2}\Delta + V(\mathbf{x}) - E_\kappa\right]\Psi_\kappa(\mathbf{x}) \to \left[-\tfrac{1}{2}\Delta - E_\kappa\right]\Psi_\kappa(\mathbf{x}) \text{ for } V(\mathbf{x}) \to 0 \quad (3.5)$$

Because of this property, all the basis functions Φ_ν have the correct asymptotic behavior. It is also desirable that all the basis functions should have turning points at approximately the same positions as the wave functions being represented so that they can contribute usefully to the synthesis of Ψ_κ. This is the case when the locus of points $\mathbf{x}$ that are solutions of

$$\beta_\nu V_0(\mathbf{x}) - E_\kappa = 0 \quad (3.6)$$

is close to the locus of points that are solutions of

$$V(\mathbf{x}) - E_\kappa = 0 \quad (3.7)$$

A third useful restriction on V_0 is to choose it as a sum of one-electron potentials, for in this case the approximate Schrödinger equation (3.4) becomes separable.

Atomic calculations using generalized Sturmians are particularly simple. If one chooses V_0 to be the nuclear attraction potential,

$$V_0(\mathbf{x}) = -\sum_{j=1}^{N} \frac{Z}{r_j} \quad (3.8)$$

then (as was shown by Goscinski in a pioneering 1968 paper [Goscinski, 1968, 2003]) an exact solution to the approximate Schrödinger equation (3.4) is given by a Slater determinant of one-electron hydrogenlike spin-orbitals:

$$\Phi_\nu(\mathbf{x}) = |\chi_{\mu_1}\chi_{\mu_2}\chi_{\mu_3}\cdots| \quad (3.9)$$

where we set $\nu \equiv (\mu_1,\ldots,\mu_N)$ and $\mu \equiv (n,l,m,m_s)$. The one-electron hydrogenlike spin-orbitals have the familiar form

$$\chi_{nlm,+1/2}(\mathbf{x}_j) = R_{nl}(r_j)Y_{lm}(\theta_j,\phi_j)\alpha(j)$$

$$\chi_{nlm,-1/2}(\mathbf{x}_j) = R_{nl}(r_j)Y_{lm}(\theta_j,\phi_j)\beta(j) \quad (3.10)$$

Table 3.1: The first few radial functions of equation (3.11).

$$R_{1,0}(r_j) = 2Q_\nu^{3/2} e^{-Q_\nu r_j}$$

$$R_{2,0}(r_j) = \frac{Q_\nu^{3/2}}{\sqrt{2}} e^{-Q_\nu r_j/2} \left(1 - \frac{Q_\nu r_j}{2}\right)$$

$$R_{2,1}(r_j) = \frac{Q_\nu^{3/2}}{2\sqrt{6}} e^{-Q_\nu r_j/2} Q_\nu r_j$$

$$\vdots \quad \vdots \quad \vdots$$

but they correspond to a weighted nuclear charge $Q_\nu = \beta_\nu Z$ that is characteristic for the configuration. In equation (3.10), Y_{lm} is a spherical harmonic and the radial part of the orbital is defined as

$$R_{nl}(r_j) = \mathcal{N}_{nl} \left(\frac{2Q_\nu r_j}{n}\right)^l e^{-Q_\nu \frac{r_j}{n}} F\left(l+1-n \middle| 2l+2 \middle| 2Q_\nu \frac{r_j}{n}\right) \tag{3.11}$$

with

$$\mathcal{N}_{nl} = \frac{2}{(2l+1)!} \left(\frac{Q_\nu}{n}\right)^{3/2} \sqrt{\frac{(l+n)!}{n(n-l-1)!}} \tag{3.12}$$

and where F is the confluent hypergeometric function ${}_1F_1$:

$$F(a|b|x) \equiv \sum_{k=0}^\infty \frac{a^{\overline{k}}}{b^{\overline{k}} k!} x^k = 1 + \frac{a}{b}\frac{x}{1!} + \frac{a(a+1)}{b(b+1)}\frac{x^2}{2!} + \cdots \tag{3.13}$$

When $a = 1+l-n$ is zero or a negative integer, the series terminates (thus fulfilling the boundary conditions of the hydrogenlike Schrödinger equation) and we have the polynomial

$$F\left(1+l-n \middle| 2l+2 \middle| 2Q_\nu \frac{r_j}{n}\right) = \sum_{k=0}^{n-l-1} \frac{(1+l-n)^{\overline{k}}}{(2l+2)^{\overline{k}} k!} \left(\frac{2Q_\nu}{n}\right)^k r_j^k \tag{3.14}$$

In order for the Goscinskian configurations shown in equation (3.9) to fulfill the approximate Schrödinger equation (3.4), one needs to choose the effective charge to be

$$Q_\nu = \beta_\nu Z = \left(\frac{-2E_\kappa}{\frac{1}{n^2} + \frac{1}{n'^2} + \cdots} \right)^{1/2} \quad (3.15)$$

The hydrogenlike one-electron spin-orbitals have the following properties: They obey the hydrogenlike wave equation

$$\left[-\frac{1}{2}\nabla_j^2 + \frac{Q_\nu^2}{2n^2} - \frac{Q_\nu}{r_j} \right] \chi_\mu(\mathbf{x_j}) = 0 \quad (3.16)$$

They are orthonormal

$$\int d\tau_j \, \chi_{\mu'}^*(\mathbf{x_j}) \chi_\mu(\mathbf{x_j}) = \delta_{\mu',\mu} \quad (3.17)$$

and they also obey the Virial Theorem, which in our case can be written in the form:

$$-\int d\tau_j \, |\chi_\mu(\mathbf{x_j})|^2 \frac{Q_\nu}{r_j} = -\frac{Q_\nu^2}{n^2} \quad (3.18)$$

Applying the kinetic energy operator to Φ_ν and using equation (3.16) for each χ_μ we have

$$\left[-\frac{1}{2} \sum_{j=1}^N \nabla_j^2 \right] \Phi_\nu(\mathbf{x}) = \left[-\left(\frac{Q_\nu^2}{2n^2} + \frac{Q_\nu^2}{2n'^2} + \cdots \right) + \left(\frac{Q_\nu}{r_1} + \frac{Q_\nu}{r_2} + \cdots \right) \right] \Phi_\nu(\mathbf{x})$$

$$= [E_\kappa - \beta_\nu V_0(\mathbf{x})] \Phi_\nu(\mathbf{x}) \quad (3.19)$$

showing that Φ_ν is indeed a solution to equation (3.4).

Solutions to the approximate Schrödinger equation (3.4) obey a potential-weighted orthonormality relation, and this can be seen in the following way: If we multiply equation (3.4) on the left by a $\Phi_{\nu'}^*$ and integrate over all space and spin coordinates, we can obtain the relationship

$$\int d\tau \, \Phi_{\nu'}^*(\mathbf{x}) \left[\frac{1}{2}\Delta + E_\kappa \right] \Phi_\nu(\mathbf{x}) = \beta_\nu \int d\tau \, \Phi_{\nu'}^*(\mathbf{x}) V_0(\mathbf{x}) \Phi_\nu(\mathbf{x}) \quad (3.20)$$

and similarly

$$\int d\tau \, \Phi_\nu^*(\mathbf{x}) \left[\frac{1}{2}\Delta + E_\kappa \right] \Phi_{\nu'}(\mathbf{x}) = \beta_{\nu'} \int d\tau \, \Phi_\nu^*(\mathbf{x}) V_0(\mathbf{x}) \Phi_{\nu'}(\mathbf{x}) \quad (3.21)$$

If we take the complex conjugate of (3.21) and subtract it from (3.20), making use of Hermiticity and assuming the β_ν's to be real, we obtain:

$$(\beta_\nu - \beta_{\nu'}) \int d\tau \, \Phi^*_{\nu'}(\mathbf{x}) V_0(\mathbf{x}) \Phi_\nu(\mathbf{x}) = 0 \qquad (3.22)$$

Thus when the two weighting factors are unequal we have

$$\int d\tau \, \Phi^*_{\nu'}(\mathbf{x}) V_0(\mathbf{x}) \Phi_\nu(\mathbf{x}) = 0 \qquad \beta_{\nu'} \neq \beta_\nu \qquad (3.23)$$

The normalization of the Goscinskian configurations can be found by making use of equations (3.15), (3.17) and (3.18) combined with the Slater-Condon rules. In this way we obtain

$$\begin{aligned}
\int d\tau \, V_0(\mathbf{x}) |\Phi_\nu(\mathbf{x})|^2 &= -\sum_{\mu \in \nu} \int d\tau_j \, |\chi_\mu(\mathbf{x}_j)|^2 \frac{Z}{r_j} \\
&= -\frac{1}{\beta_\nu} \sum_{\mu \in \nu} \int d\tau_j \, |\chi_\mu(\mathbf{x}_j)|^2 \frac{Q_\nu}{r_j} \\
&= -\frac{Q_\nu^2}{\beta_\nu} \sum_{\mu \in \nu} \frac{1}{n^2} = \frac{2E_\kappa}{\beta_\nu}
\end{aligned} \qquad (3.24)$$

Combining equations (3.23) and (3.24), we obtain a potential-weighted orthonormality relation for the Goscinskian configurations:

$$\int d\tau \, \Phi^*_{\nu'}(\mathbf{x}) V_0(\mathbf{x}) \Phi_\nu(\mathbf{x}) = \delta_{\nu',\nu} \frac{2E_\kappa}{\beta_\nu} \qquad (3.25)$$

Something needs to be added concerning equation (3.25): Since the index ν stands for the whole set of quantum numbers of a configuration in our basis set, there are many cases for which $\nu' \neq \nu$ but $\beta_{\nu'} = \beta_\nu$. For such cases, potential-weighted orthogonality of the two configurations does not follow from (3.23). However, when the two β's are equal,

$$Q_\nu = \beta_\nu Z = \beta_{\nu'} Z = Q_{\nu'}$$

and therefore

$$\mathcal{R}_\nu = \frac{p_\kappa}{Q_\nu} = \frac{p_\kappa}{Q_{\nu'}} = \mathcal{R}_{\nu'}$$

Since $Q_\nu = Q_{\nu'}$ we have radial orthonormality of the atomic orbitals, and therefore the usual Slater-Condon rules hold.

In order that $\mathcal{R}_\nu = \mathcal{R}_{\nu'}$ when $\nu \neq \nu'$, at least two of the μ's occurring in ν must differ from the μ's occurring in ν'. From the Slater-Condon rules it then follows that the inner product shown in (3.23) is zero whenever

$\nu \neq \nu'$. Thus the potential-weighted orthonormality relation (3.25) holds in all cases.

3.2 Derivation of the secular equations

We are now in a position to derive the generalized Sturmian secular equations appropriate for atoms and ions. In this derivation, it will be convenient to introduce the variables

$$p_\kappa \equiv \sqrt{-2E_\kappa} \qquad (3.26)$$

and

$$\mathcal{R}_\nu \equiv \sqrt{\frac{1}{n^2} + \frac{1}{n'^2} + \cdots} \qquad (3.27)$$

In equation (3.27), the n's are the principal quantum numbers of the hydrogenlike spin-orbitals occurring in the Goscinskian configurations Φ_ν (3.9). In terms of these two new variables, equation (3.15) takes on the simple form:

$$Q_\nu = \beta_\nu Z = \frac{p_\kappa}{\mathcal{R}_\nu} \qquad (3.28)$$

From equation (3.26) it follows that

$$E_\kappa = -\frac{p_\kappa^2}{2} \qquad (3.29)$$

If we neglect the motion of the nucleus, spin-orbit coupling and spin-spin coupling, the potential that appears in the non-relativistic Schrödinger equation of an atom or ion without external fields (equation (3.1)) is

$$V(\mathbf{x}) = V_0(\mathbf{x}) + V'(\mathbf{x}) = -\sum_{j=1}^{N} \frac{Z}{r_j} + \sum_{j>i}^{N} \sum_{i=1}^{N} \frac{1}{r_{ij}} \qquad (3.30)$$

where V_0 and V' respectively represent nuclear attraction and interelectron repulsion. We next introduce the matrices

$$T^0_{\nu',\nu} \equiv -\frac{1}{p_\kappa} \int d\tau \ \Phi^*_{\nu'}(\mathbf{x}) V_0(\mathbf{x}) \Phi_\nu(\mathbf{x}) \qquad (3.31)$$

and

$$T'_{\nu',\nu} \equiv -\frac{1}{p_\kappa} \int d\tau \ \Phi^*_{\nu'}(\mathbf{x}) V'(\mathbf{x}) \Phi_\nu(\mathbf{x}) \qquad (3.32)$$

From the potential-weighted orthogonality relation and normalization (3.25) and from equation (3.28) it follows that

$$T^0_{\nu',\nu} = \delta_{\nu'\nu} Z\mathcal{R}_\nu \qquad (3.33)$$

Thus the nuclear attraction matrix $T^0_{\nu',\nu}$ defined by equation (3.31) is independent of p_κ. It turns out that $T'_{\nu',\nu}$ is also independent of p_κ, as is shown in Appendix A.

We now substitute the superposition (3.3) into the Schrödinger equation (3.1):

$$\sum_\nu \left[-\frac{1}{2}\Delta + V(\mathbf{x}) - E_\kappa \right] \Phi_\nu(\mathbf{x}) B_{\nu,\kappa} = 0 \qquad (3.34)$$

Since each of the basis functions Φ_ν obeys the approximate Schrödinger equation (3.4), equation (3.34) can be written in the form

$$\sum_\nu \left[V(\mathbf{x}) - \beta_\nu V_0(\mathbf{x}) \right] \Phi_\nu(\mathbf{x}) B_{\nu,\kappa} = 0 \qquad (3.35)$$

If we multiply (3.35) from the left by a conjugate function from the basis set and integrate over space and spin coordinates, we obtain

$$\sum_\nu \int d\tau\, \Phi^*_{\nu'}(\mathbf{x}) \left[V(\mathbf{x}) - \beta_\nu V_0(\mathbf{x}) \right] \Phi_\nu(\mathbf{x}) B_{\nu,\kappa} = 0 \qquad (3.36)$$

With the help of equations (3.30), (3.32) and (3.33), this becomes

$$\sum_\nu \left[-p_\kappa \delta_{\nu',\nu} Z\mathcal{R}_\nu - p_\kappa T'_{\nu',\nu} + \beta_\nu p_\kappa \delta_{\nu',\nu} Z\mathcal{R}_\nu \right] B_{\nu,\kappa} = 0 \qquad (3.37)$$

Finally, making use of the fact that

$$\beta_\nu Z\mathcal{R}_\nu = p_\kappa \qquad (3.38)$$

changing signs, and dividing by p_κ, we obtain:

$$\sum_\nu \left[\delta_{\nu',\nu} Z\mathcal{R}_\nu + T'_{\nu',\nu} - p_\kappa \delta_{\nu',\nu} \right] B_{\nu,\kappa} = 0 \qquad (3.39)$$

Equation (3.39) differs in several remarkable respects from the conventional set of secular equations that we would obtain by diagonalizing the Hamiltonian of an atom or ion:

(1) The kinetic energy term has vanished.
(2) The nuclear attraction matrix is diagonal and energy-independent.
(3) The interelectron repulsion matrix is energy-independent.

(4) The roots are not energies but values of the parameter p_κ, proportional to the square roots of the binding energies (equation (3.26)).

Because the effective nuclear charges that characterize the basis functions depend on p_κ, the basis set is not completely known before the secular equations have been solved. Only the form of the basis functions is known, but not their scale. Solving the secular equations gives us at one stroke a spectrum of energies, and a near-optimal set of basis functions for representing the corresponding states. The Slater-exponents that appear in the hydrogenlike orbitals of the Goscinskian configurations turn out to be almost exactly those that would have been found by a variational optimization of the basis. In the generalized Sturmian method, we are able to dispense with the time-consuming variational optimization that would otherwise be needed to find the optimum basis for representing each excited state. This characteristic of the generalized Sturmian method makes it especially suitable for calculating large numbers of excited states of few-electron atoms.

3.3 Symmetry-adapted basis sets for the 2-electron isoelectronic series

For the 2-electron isoelectronic series, it is easy to construct symmetry-adapted basis functions of the Russell-Saunders type by means of Clebsch-Gordan coefficients. Such basis functions are simultaneous eigenfunctions of the total orbital angular momentum operator L^2, its z-component, L_z, the total spin operator, S^2, and its z-component S_z. If we let η_j stand for such a Russell-Saunders-type basis function, then the equation representing it as a superposition of primitive Goscinskian configurations can be written in the form:

$$\eta_j(\mathbf{x}) = \sum_\nu \Phi_\nu(\mathbf{x}) C_{\nu,j} \qquad (3.40)$$

where the coefficients in the superposition are products of Clebsch-Gordan coefficients:

$$C_{\nu,j} \equiv \begin{pmatrix} l & l' & L \\ m & m' & M \end{pmatrix} \begin{pmatrix} \frac{1}{2} & \frac{1}{2} & S \\ m_s & m'_s & M_S \end{pmatrix} \qquad (3.41)$$

In equations (3.40) and (3.41), the indices ν and j refer to the sets of quantum numbers

$$\nu \equiv (n, l, m, m_s; n', l', m', m'_s) \qquad (3.42)$$

and
$$j \equiv (n, l; n', l'; L, M; S, M_S) \tag{3.43}$$
which respectively characterize the primitive configurations
$$\Phi_\nu(\mathbf{x}) \equiv |\chi_{n,l,m,m_s} \chi_{n',l',m',m'_s}| \tag{3.44}$$
and the symmetry-adapted basis functions η_j. Since C is composed of Clebsch-Gordan coefficients, we have
$$C^\dagger C = I \tag{3.45}$$

We now transform the nuclear attraction matrix to the symmetry-adapted representation:
$$\tilde{T}^0 = C^\dagger T^0 C \tag{3.46}$$
Looking at equations (3.33), (3.45), (3.46) and at the definition of $\mathcal{R}_\nu$ (3.27), we can see that the nuclear attraction matrix will also be diagonal in the new representation
$$\tilde{T}^0{}_{j',j} = \delta_{j',j} Z \mathcal{R}_j \tag{3.47}$$
since the Clebsch-Gordan coefficients can only mix primitive configurations belonging to the same value of $\mathcal{R}_\nu$. We next transform the interelectron repulsion matrix defined by equation (3.32):
$$\tilde{T}' = C^\dagger T' C \tag{3.48}$$
This gives us the symmetry-adapted secular equations
$$\sum_{j \in I} \left[\delta_{j',j} Z \mathcal{R}_j + \tilde{T}'{}_{j',j} - p_\kappa \delta_{j',j} \right] \tilde{B}_{j,\kappa} = 0 \tag{3.49}$$
where I is the set of all values of $j = (n, l, n', l', L, M, S, M_S)$ for which the last four quantum numbers are the chosen values of L, M, S and M_S. The coefficients $B_{j,\kappa}$ which result from the solution of (3.49) allow us to express the wave function of the atom or ion in its various states as superpositions of symmetry-adapted basis functions:
$$\Psi_\kappa(\mathbf{x}) = \sum_j \eta_j(\mathbf{x}) \tilde{B}_{j,\kappa} \tag{3.50}$$

The energies are obtained by substituting the spectrum of p_κ values into (3.29). In this way one can obtain both singly-excited and doubly-excited states that agree well with experimental values [NIFS database], [NIST database], as shown in Tables 3.2-3.4.

Table 3.2: ^1S excited state energies (in Hartrees) for the 2-electron isoelectronic series. The basis set used consisted of 40 generalized Sturmians of the Goscinski type. Experimental values are taken from the NIST tables [NIST database] (http://physics.nist.gov/asd). Discrepancies between calculated and experimental energies for the heavier ions are due mainly to relativistic effects, which are discussed in the text.

	He	Li^+	Be^{2+}	B^{3+}	C^{4+}	N^{5+}
$1s2s\ ^1S$	−2.1429	−5.0329	−9.1730	−14.564	−21.206	−29.098
expt.	−2.1458	−5.0410	−9.1860	−14.582	−21.230	−29.131
$1s3s\ ^1S$	−2.0603	−4.7297	−8.5099	−13.402	−19.406	−26.521
expt.	−2.0611	−4.7339	−8.5183	−13.415	−19.425	−26.548
$1s4s\ ^1S$	−2.0332	−4.6276	−8.2837	−13.003	−18.785	−25.629
expt.	−2.0334	−4.6299	−8.2891			−25.654
$1s5s\ ^1S$	−2.0210	−4.5811	−8.1806	−12.820	−18.500	−25.220
expt.	−2.0210	−4.5825				−25.241
$1s6s\ ^1S$	−2.0144	−4.5562	−8.1250	−12.721	−18.346	−24.998
expt.	−2.0144	−4.5571				
$1s7s\ ^1S$	−2.0105	−4.5412	−8.0917	−12.662	−18.253	−24.865
expt.	−2.0104	−4.5418				
$1s8s\ ^1S$	−2.0080	−4.5315	−8.0701	−12.624	−18.194	−24.779
expt.	−2.0079					
$1s9s\ ^1S$	−2.0063	−4.5248	−8.0554	−12.598	−18.153	−24.720
expt.	−2.0062					
$1s10s\ ^1S$	−2.0051	−4.5201	−8.0449	−12.579	−18.124	−24.678
expt.	−2.0050					
$1s11s\ ^1S$	−2.0042	−4.5166	−8.0371	−12.566	−18.102	−24.647
expt.	−2.0041					
$1s12s\ ^1S$	−2.0034	−4.5140	−8.0312	−12.555	−18.086	−24.624
expt.	−2.0034					

Table 3.3: ^1D excited state energies for the 2-electron isoelectronic series, compared with experimental values taken from the NIST tables [NIST database].

	He	Li^+	Be^{2+}	B^{3+}	C^{4+}	N^{5+}
$1s3d\ ^1D$	−2.0555	−4.7218	−8.4990	−13.388	−19.387	−26.498
$expt.$	−2.0554	−4.7225	−8.5012	−13.392	−19.396	−26.514
$1s4d\ ^1D$	−2.0312	−4.6246	−8.2801	−12.998	−18.779	−25.622
$expt.$	−2.0311	−4.6252	−8.2824	−13.003	−18.788	−25.639
$1s5d\ ^1D$	−2.0200	−4.5797	−8.1790	−12.818	−18.497	−25.217
$expt.$	−2.0198	−4.5801	−8.1807		−18.507	−25.234
$1s6d\ ^1D$	−2.0139	−4.5554	−8.1242	−12.721	−18.345	−24.997
$expt.$	−2.0137	−4.5557			−18.354	
$1s7d\ ^1D$	−2.0102	−4.5407	−8.0912	−12.662	−18.253	−24.865
$expt.$	−2.0100	−4.5409			−18.262	
$1s8d\ ^1D$	−2.0078	−4.5311	−8.0699	−12.624	−18.194	−24.779
$expt.$	−2.0076	−4.5314			−18.202	
$1s9d\ ^1D$	−2.0062	−4.5246	−8.0552	−12.598	−18.153	−24.720
$expt.$	−2.0060					
$1s10d\ ^1D$	−2.0050	−4.5199	−8.0447	−12.579	−18.124	−24.678
$expt.$	−2.0048					
$1s11d\ ^1D$	−2.0041	−4.5165	−8.0370	−12.566	−18.102	−24.647
$expt.$	−2.0035					
$1s12d\ ^1D$	−2.0032	−4.5139	−8.0311	−12.555	−18.086	−24.624
$expt.$	−2.0033					

Table 3.4: Doubly-excited (autoionizing) ^{3}S states of the 2-electron isoelectronic series with $n=2$ and $n'=3, 4, 5$. See also Figure 3.1.

	2s3s ^{3}S	2p3p ^{3}S	2s4s ^{3}S	2p4p ^{3}S	2s5s ^{3}S	2p5p ^{3}S
C^{4+}	−6.1198	−5.9299	−5.3750	−5.2906	−5.0469	−5.0028
expt.	−6.1221	−5.9299				
N^{5+}	−8.4020	−8.1756	−7.3629	−7.2614	−6.9017	−6.8484
expt.	−8.4058	−8.1768				
O^{6+}	−11.0452	−10.7824	−9.6633	−9.5447	−9.0464	−8.9839
expt.	−11.0545	−10.7869				
F^{7+}	−14.0496	−13.7504	−12.2762	−12.1405	−11.4812	−11.4095
expt.	−14.0616	−13.7556				
Ne^{8+}	−17.4151	−17.0795	−15.2017	−15.0488	−14.2059	−14.1251
expt.	−17.4350	−17.0905				
Na^{9+}	−21.1417	−20.7697	−18.4396	−18.2696	−17.2207	−17.1307
expt.	−21.1749	−20.7888				
Mg^{10+}	−25.2295	−24.8210	−21.9900	−21.8030	−20.5255	−20.4264
expt.	−25.2767	−24.8504				
Al^{11+}	−29.6783	−29.2334	−25.8529	−25.6488	−24.1203	−24.0121
expt.	−29.7450	−29.2761				
Si^{12+}	−34.4883	−34.0069	−30.0284	−29.8072	−28.0050	−27.8877
expt.	−34.5809	−34.0654				
P^{13+}	−39.6593	−39.1416	−34.5163	−34.2780	−32.1798	−32.0534
expt.	−39.7833	−39.2138				
S^{14+}	−45.1915	−44.6373	−39.3167	−39.0614	−36.6446	−36.5091
expt.	−45.3567	−44.7298				
Cl^{15+}	−51.0848	−50.4942	−44.4296	−44.1573	−41.3994	−41.2548
expt.	−51.2969	−50.6131				
Ar^{16+}	−57.3392	−56.7122	−49.8551	−49.5656	−46.4442	−46.2905
expt.	−46.3829	−45.6295				
K^{17+}	−63.9547	−63.2913	−55.5930	−55.2865	−51.7790	−51.6162
expt.	−64.2961	−63.4726				
Ca^{18+}	−70.9313	−70.2315	−61.6434	−61.3199	−57.4038	−57.2320
expt.	−71.3560	−70.4479				
Sc^{19+}	−78.2690	−77.5329	−68.0064	−67.6657	−63.3186	−63.1377
expt.	−78.7497	−77.7470				
Ti^{20+}	−85.9678	−85.1953	−74.6818	−74.3241	−69.5234	−69.3334
expt.		−85.4971				

3.4 The large-Z approximation

When nuclear attraction dominates completely over interelectron repulsion, the term $T'_{\nu',\nu}$ can be neglected in (3.39), and with the help of (3.29) we obtain

$$\frac{E_\kappa}{Z^2} \to -\frac{1}{2}\mathcal{R}_\nu^2 = -\frac{1}{2n^2} - \frac{1}{2n'^2} - \frac{1}{2n''^2} - \cdots \qquad (3.51)$$

In other words, if interelectron repulsion is totally neglected, the energy levels of an N-electron atom correspond to those of a set of N entirely independent electrons moving in the attractive potential of a nucleus with charge Z. We can introduce the name "$\mathcal{R}$-block" for the set of primitive configurations that are degenerate when interelectron repulsion is neglected. Even when Z is large, interelectron repulsion hybridizes the set of degenerate primitive configurations belonging to a particular value of $\mathcal{R}_\nu$ and very slightly removes the degeneracy. The large-Z approximation consists of neglecting all configurations outside the $\mathcal{R}$-block. In other words, for large Z, the interelectron repulsion matrix elements linking different $\mathcal{R}$-blocks are neglected, and equation (3.39) is solved one block at a time. For each $\mathcal{R}$-block, the term $\delta_{\nu',\nu} Z \mathcal{R}_\nu$ is a multiple of the unit matrix. From linearity, it follows that adding any multiple of the unit matrix to another matrix does not change its eigenvectors, while all its roots are shifted by a constant amount. Therefore in the large-Z approximation we can find solutions for (3.39) by diagonalizing the interelectron repulsion matrix for a particular $\mathcal{R}$-block, i.e. by solving

$$\sum_\nu {}' \left[T'_{\nu',\nu} - \lambda_\kappa \delta_{\nu',\nu} \right] B_{\nu,\kappa} = 0 \qquad (3.52)$$

The sum in equation (3.52) is taken over all values of ν that correspond to a particular value of $\mathcal{R}_\nu$, i.e. to a particular set of principal quantum numbers $n, n', \ldots$. Thus in the large-Z approximation, the roots to the secular equations (3.39) are the same as those of (3.52), except that they are shifted by a constant amount, $Z\mathcal{R}_\nu$.

$$p_\kappa = Z\mathcal{R}_\nu + \lambda_\kappa \qquad (3.53)$$

Then with the help of (3.29) we obtain

$$E_\kappa \approx -\frac{1}{2}(Z\mathcal{R}_\nu + \lambda_\kappa)^2 \qquad (3.54)$$

Because of the minus sign in the definition of $T'_{\nu',\nu}$, the roots of the interelectron repulsion matrix are always negative. Thus, as we would expect, the effect of interelectron repulsion is to reduce the binding energy of a state. This point can be made more explicit by writing

$$E_\kappa \approx -\frac{1}{2}(Z\mathcal{R}_\nu - |\lambda_\kappa|)^2 \tag{3.55}$$

The true energy is only approximately equal to the expression on the right of (3.55) because of the restriction of the basis set to a single $\mathcal{R}$-block, but as the nuclear charge increases in an isoelectronic series, the approximation improves.

To illustrate the large-Z approximation, let us consider doubly-excited (autoionizing) states of the 2-electron isoelectronic series. For example, we can diagonalize the interelectron repulsion matrix for the $\mathcal{R}$-block characterized by the principal quantum numbers $n = 2$ and $n' = 3$. If we include all of the primitive configurations with these principle quantum numbers, the block will give us a 144×144 matrix to diagonalize in equation (3.52), since $4n^2 n'^2 = 144$. However, if we restrict the primitive configurations to those characterized by $M_S = 1$, the block of $T'_{\nu',\nu}$ is only 36×36. When we diagonalize the interelectron repulsion matrix for this smaller block, we obtain 36 roots, λ_κ, and it can be observed that each of the roots is $(2L+1)$-fold degenerate. (If the full 144×144 $\mathcal{R}$-block had been used, each of the roots would have been $(2L + 1) \times (2S + 1)$-fold degenerate.)

Analysis of the eigenvectors then shows that they are symmetry-adapted basis functions of the Russell-Saunders type, i.e. they are simultaneous eigenfunctions of the operators L^2, L_z, S^2 and S_z. In fact they are exactly the same symmetry-adapted basis functions that would have been obtained if we had used Clebsch-Gordan coefficients in the manner described in the previous section. This is because the invariance of V' under rotation implies that the four operators just mentioned all commute with V', and therefore simultaneous eigenfunctions of all five operators can be obtained.

Although we know that it is possible to obtain simultaneous eigenfunctions of all five operators, it does not follow that these simultaneous eigenfunctions will be obtained by diagonalizing the interelectron repulsion matrix. In order to produce simultaneous eigenfunctions of L^2, L_z, S^2 and S_z, it is necessary to remove the degeneracy very slightly by adding an extremely small external perturbation of the form

$$V_{ext} = \epsilon_L L_z + \epsilon_S S_z \tag{3.56}$$

where, for example, $\epsilon_L = 10^{-10}$ and $\epsilon_S = 10^{-12}$.

Looking at the first column of Table 3.5, we can see that the root with smallest absolute magnitude, $|\lambda_\kappa| = .108252$, corresponds to a ^3S state. There is another root corresponding to ^3S symmetry with magnitude $|\lambda_\kappa| = .168814$. Inserting these two roots and the appropriate value of $\mathcal{R}_\nu$ into

ATOMIC SPECTRA

Table 3.5: Eigenvalues of the 2-electron interelectron repulsion matrix $T'_{\nu',\nu}$ for $S=1$, $M_S=1$, $n=2$ and $n'=3, 4, 5$.

$n'=3$ $\|\lambda_\kappa\|$	term	$n'=4$ $\|\lambda_\kappa\|$	term	$n'=5$ $\|\lambda_\kappa\|$	term
.108252	^{3}S	.077484	^{3}S	.056075	^{3}S
.134734	^{3}P	.087582	^{3}P	.065019	^{3}P
.135408	^{3}D	.090845	^{3}D	.061128	^{3}P
.138421	^{3}P	.093401	^{3}P	.063370	^{3}D
.155155	^{3}F	.099235	^{3}F	.067758	^{3}F
.160439	^{3}P	.099991	^{3}P	.067934	^{3}P
.165613	^{3}D	.104253	^{3}D	.070494	^{3}D
.168814	^{3}S	.106271	^{3}D	.071269	^{3}D
.173917	^{3}D	.107976	^{3}S	.072413	^{3}F
.186893	^{3}P	.108188	^{3}F	.072857	^{3}S
		.111210	^{3}G	.073295	^{3}G
		.111264	^{3}F	.073588	^{3}G
		.113313	^{3}P	.073920	^{3}F
		.114381	^{3}D	.074306	^{3}G
				.074578	^{3}H
				.074963	^{3}F
				.075173	^{3}P
				.075545	^{3}D

equation (3.55), we obtain respectively

$$E_\kappa \approx -\frac{1}{2}\left(Z\sqrt{\frac{1}{2^2}+\frac{1}{3^2}} - .108252\right)^2 \tag{3.57}$$

and

$$E_\kappa \approx -\frac{1}{2}\left(Z\sqrt{\frac{1}{2^2}+\frac{1}{3^2}} - .168814\right)^2 \tag{3.58}$$

These two equations correspond to the lowest two curves shown in Figure 3.1. This figure shows the ratio E_κ/Z^2 as a function of Z for doubly-excited (autoionizing) ^{3}S states of the 2-electron isoelectronic series. For the states shown in the figure, the lowest principal quantum number has the value $n = 2$, while the other principal quantum number takes on the values $n' = 3$, $n' = 4$ and $n' = 5$. As Z grows, the ratios E_κ/Z^2 approach the Z-independent values given by equation (3.51).

The third-lowest curve in the figure is represented by

$$E_\kappa \approx -\frac{1}{2}\left(Z\sqrt{\frac{1}{2^2} + \frac{1}{4^2}} - .077484\right)^2 \tag{3.59}$$

while the fourth-lowest is given by

$$E_\kappa \approx -\frac{1}{2}\left(Z\sqrt{\frac{1}{2^2} + \frac{1}{4^2}} - .107976\right)^2 \tag{3.60}$$

The roots appearing in equations (3.59) and (3.60) can be found in the third column of Table 3.5. The dots in Figure 3.1 correspond to energies obtained from the full basis, and, on the scale of the figure, they are indistinguishable from the experimental values shown in Table 3.4. As can be seen from the figure, the large-Z approximation becomes highly accurate for values of Z greater than about 20, and it retains some qualitative validity for lower values of Z. However we must remember that we are looking at doubly-excited states of the 2-electron isoelectronic series. For these states the electrons can be visualized as being far apart, and interelectron repulsion plays a smaller role than it does for less highly excited states of atoms and ions. Relativistic effects [Akhiezer and Berestetskii, 1965], [Avery, 1976], [Avery, 2000] are also much less important than they would be if the electrons were in orbitals close to the nuclei.

Equations (3.52) and (3.55) can be applied to any state whatever of an atom or ion in the large-Z approximation. To illustrate this point, we can think of the lowest carbon-like $\mathcal{R}$-block with

$$\mathcal{R}_\nu \equiv \sqrt{\frac{2}{1^2} + \frac{4}{2^2}} = \sqrt{3} \tag{3.61}$$

In other words, we consider the $\mathcal{R}$-block corresponding to a 6-electron configuration with 2 electrons in the $n = 1$ shell and 4 electrons in the $n = 2$ shell.

Because of the Pauli principle the full $\mathcal{R}$-block has a much smaller dimension than $2n^2 \times 2n'^2 \times 2n''^2 \times \cdots$. In fact, the carbon-like $\mathcal{R}$-block gives us a 70×70 matrix to diagonalize in equation (3.52). The dimension of the $\mathcal{R}$-block is in fact given by the binomial coefficient

$$\binom{8}{4} = \frac{8!}{4!(8-4)!} = 70 \tag{3.62}$$

because this is the number of ways to select 4 different spin-orbitals from the 8 available one-electron hydrogenlike functions of the $n = 2$ shell, while there is only one way of putting the 2 remaining electrons into the $n = 1$ shell.

The roots and eigenvectors of the interelectron repulsion matrix corresponding to the carbon-like $\mathcal{R}$-block with $\mathcal{R}_\nu = \sqrt{3}$ are shown in Table 3.6. Each root corresponds to a particular value of L and S and the roots are $(2L+1) \times (2S+1)$-fold degenerate. The reader can verify that the sum of the degeneracies is equal to 70, the dimension of the $\mathcal{R}$-block. Substituting the root with the smallest numerical value into equation (3.55), we obtain

$$E_\kappa \approx -\frac{1}{2}\left(Z\sqrt{\frac{2}{1^2} + \frac{4}{2^2}} - 1.88151\right)^2 \tag{3.63}$$

for the energy of the ground state of the 6-electron isoelectronic series in the large-Z approximation. The large-Z approximation predicts that the ground state state will be a ^{3}P state, and this agrees with experiment. Equation (3.55) and Table 3.6 also predict that the next few higher states will be respectively ^{1}D, ^{1}S, ^{5}S, ^{3}D and ^{3}P, and this sequence is confirmed by experiment when Z is large. Because interelectron repulsion plays a much larger role in the ground states of the 6-electron isoelectronic series than in the doubly-excited states of the 2-electron isoelectronic series, the large-Z approximation has less accuracy here than in the case shown in Figure 3.1. Nevertheless the approximation has some qualitative validity, and it improves with increasing Z.

The right-hand column of Table 3.6 shows the composition of the eigenvectors with $M = L$ and $M_S = S$. These are the simplest of the eigenvectors. In order to make the eigenvectors separate cleanly into simultaneous eigenfunctions of the operators V', L^2, L_z S^2 and S_z, it was necessary to add an extremely small external perturbation as in equation (3.56).

Table 3.6: Eigenvalues of $T'_{\nu',\nu}$ for the carbon-like $\mathcal{R}_\nu = \sqrt{3}$ block.

$|\lambda_\kappa|$	term	degen.	configuration
1.88151	^{3}P	9	$.994467(1s)^2(2s)^2(2p)^2 + .105047(1s)^2(2p)^4$
1.89369	^{1}D	5	$.994467(1s)^2(2s)^2(2p)^2 - .105047(1s)^2(2p)^4$
1.90681	^{1}S	1	$.979686(1s)^2(2s)^2(2p)^2 + .200537(1s)^2(2p)^4$
1.91623	^{5}S	5	$(1s)^2(2s)(2p)^3$
1.95141	^{3}D	15	$(1s)^2(2s)(2p)^3$
1.96359	^{3}P	9	$(1s)^2(2s)(2p)^3$
1.98389	^{3}S	3	$(1s)^2(2s)(2p)^3$
1.98524	^{1}D	5	$(1s)^2(2s)(2p)^3$
1.99742	^{1}P	3	$(1s)^2(2s)(2p)^3$
2.04342	^{3}P	9	$.105047(1s)^2(2s)^2(2p)^2 - .994467(1s)^2(2p)^4$
2.05560	^{1}D	5	$.105047(1s)^2(2s)^2(2p)^2 + .994467(1s)^2(2p)^4$
2.07900	^{1}S	1	$.200537(1s)^2(2s)^2(2p)^2 - .979686(1s)^2(2p)^4$

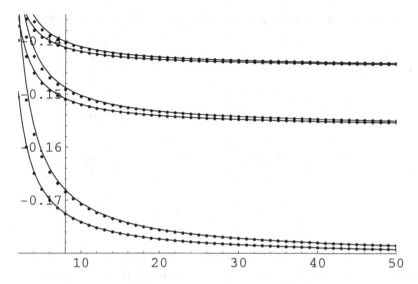

Fig. 3.1 The large-Z approximation for E_κ/Z^2 as a function of Z (smooth curves) is compared in this figure with values calculated using a fuller basis set (dots). The curves in the figure, starting from the bottom, correspond to 2s3s ^{3}S states, equation (3.57), 2p3p ^{3}S states, equation (3.58), and so on. See also Table 3.4.

It is also interesting to consider the lowest $\mathcal{R}$-blocks for the He-like, Li-like, Be-like and B-like isoelectronic series. These have $\mathcal{R}_\nu$ values and dimensions given respectively by

$$\text{He-like}: \quad \mathcal{R}_\nu = \sqrt{\frac{2}{1^2}} = \sqrt{2} \qquad \binom{8}{0} = \frac{8!}{0!(8-0)!} = 1$$

$$\text{Li-like}: \quad \mathcal{R}_\nu = \sqrt{\frac{2}{1^2} + \frac{1}{2^2}} = \frac{3}{2} \qquad \binom{8}{1} = \frac{8!}{1!(8-1)!} = 8$$

$$\text{Be-like}: \quad \mathcal{R}_\nu = \sqrt{\frac{2}{1^2} + \frac{2}{2^2}} = \frac{\sqrt{10}}{2} \qquad \binom{8}{2} = \frac{8!}{2!(8-2)!} = 28 \quad (3.64)$$

$$\text{B-like}: \quad \mathcal{R}_\nu = \sqrt{\frac{2}{1^2} + \frac{3}{2^2}} = \frac{\sqrt{11}}{2} \qquad \binom{8}{3} = \frac{3!}{3!(8-3)!} = 56$$

The roots $|\lambda_\kappa|$ for these $\mathcal{R}$-blocks are shown in Table 3.4. Similarly, the lowest $\mathcal{R}$-blocks of the interelectron repulsion matrix for the N-like, O-like,

F-like and Ne-like isoelectronic series are characterized respectively by

$$\text{N-like}: \quad \mathcal{R}_\nu = \sqrt{\frac{2}{1^2} + \frac{5}{2^2}} = \frac{\sqrt{13}}{2} \qquad \binom{8}{5} = \frac{8!}{5!(8-5)!} = 56$$

$$\text{O-like}: \quad \mathcal{R}_\nu = \sqrt{\frac{2}{1^2} + \frac{6}{2^2}} = \frac{\sqrt{14}}{2} \qquad \binom{8}{6} = \frac{8!}{6!(8-6)!} = 28$$

$$\text{F-like}: \quad \mathcal{R}_\nu = \sqrt{\frac{2}{1^2} + \frac{7}{2^2}} = \frac{\sqrt{15}}{2} \qquad \binom{8}{7} = \frac{8!}{7!(8-7)!} = 8 \qquad (3.65)$$

$$\text{Ne-like}: \quad \mathcal{R}_\nu = \sqrt{\frac{2}{1^2} + \frac{8}{2^2}} = 2 \qquad \binom{8}{8} = \frac{8!}{8!(8-8)!} = 1$$

The roots and symmetries obtained by diagonalizing these $\mathcal{R}$-blocks are shown in Table 3.8. The solid lines in Figure 3.2 show the ground state energies of the heliumlike and lithiumlike isoelectronic series calculated in the large-Z approximation, using the equations

$$\text{He-like}: \quad E_\kappa \simeq -\tfrac{1}{2}\left(Z\sqrt{\frac{2}{1^2}} - 0.441942\right)^2$$

$$(3.66)$$

$$\text{Li-like}: \quad E_\kappa \simeq -\tfrac{1}{2}\left(Z\sqrt{\frac{2}{1^2} + \frac{1}{2^2}} - 0.68180\right)^2$$

The roots of the ground state $\mathcal{R}$-block of the interelectron repulsion matrix for these two series are taken from Table 3.7, and they correspond respectively to ^1S and ^2S states. The dots in the figure are experimental values taken from the National Institute of Standards and Technology spectroscopic tables [NIST database]. It can be seen from the figure that the large-Z approximation differs from the experimental values as the atomic number increases. This is because with increasing Z the large-Z approximation approaches an exact solution to the non-relativistic Schrödinger equation (3.1). However, as Z increases, the core electrons move with velocities approaching that of light, and relativistic effects become important. In Figure 3.1 and Table 3.4, relativistic effects are less important, because for doubly excited states of the 2-electron isoelectronic series, both electrons occupy orbitals far from the nucleus.

Figure 3.3 shows the experimental ground state energies compared with the expressions in equation (3.66) multiplied by a relativistic correction

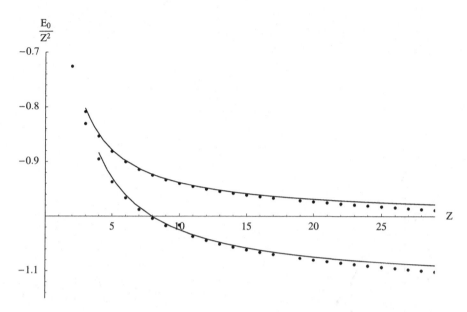

Fig. 3.2 The large-Z approximations $E_\kappa/Z^2 \simeq -(Z\mathcal{R}_\nu - |\lambda_\kappa|)^2/(2Z^2)$ (smooth curves) are compared to experimental values from NIST (dots) for the ground states of the heliumlike and lithiumlike isoelectronic series. Failure of the large-Z approximation to match experimental ground states as Z increases is attributed to neglecting relativistic effects, the role of which increases as the nuclei become heavier.

factor $f_\nu(Z)$. The factor used was the ratio between the relativistic energy of the configuration completely neglecting interelectron repulsion and the non-relativistic energy $-\frac{1}{2}(Z\mathcal{R}_\nu)^2$ also neglecting interelectron repulsion. The relativistic 1-electron energies, including the rest energy and expressed in Hartrees, are given by

$$\epsilon_\mu = c^2 \left[1 + \left(\frac{Z}{c(\gamma + n - |j + 1/2|)} \right)^2 \right]^{-1/2} \tag{3.67}$$

$$\gamma \equiv \sqrt{\left(j + \frac{1}{2}\right)^2 - \left(\frac{Z}{c}\right)^2} \qquad c = 137.036 \tag{3.68}$$

where j is the total angular momentum (orbital plus spin) of a single electron, i.e. $l \pm \frac{1}{2}$. Chapter 7 will discuss relativistic effects in much greater detail.

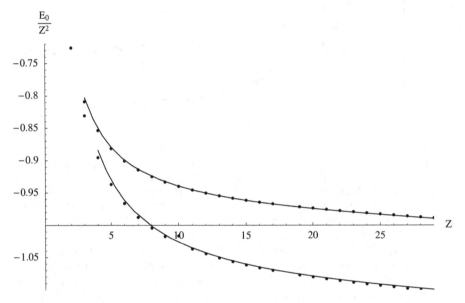

Fig. 3.3 This figure shows the large-Z approximation for the heliumlike and lithiumlike isoelectronic series ground states crudely corrected for relativistic effects. The smooth lines show $E_\kappa/Z^2 \simeq -f_\nu(Z)(Z\mathcal{R}_\nu - |\lambda_\kappa|)^2/(2Z^2)$. Here $f_\nu(Z)$ is the ratio between the energies with and without relativistic effects when one neglects interelectron repulsion. The dots represent experimental values. The improved agreement at large Z suggests that deviations in the previous figure can be attributed to relativistic effects.

Table 3.7: Roots of the lowest $\mathcal{R}$-block of the interelectron repulsion matrix for the He-like, Li-like, Be-like and B-like isoelectronic series.

He-like $\|\lambda_\kappa\|$	term	Li-like $\|\lambda_\kappa\|$	term	Be-like $\|\lambda_\kappa\|$	term	B-like $\|\lambda_\kappa\|$	term
0.441942	^{1}S	0.681870	^{2}S	0.986172	^{1}S	1.40355	^{2}P
		0.729017	^{2}P	1.02720	^{3}P	1.44095	^{4}P
				1.06426	^{1}P	1.47134	^{2}D
				1.09169	^{3}P	1.49042	^{2}S
				1.10503	^{1}D	1.49395	^{2}P
				1.13246	^{1}S	1.52129	^{4}S
						1.54037	^{2}D
						1.55726	^{2}P

Table 3.8: Roots of the lowest $\mathcal{R}$-block of $T'_{\nu',\nu}$ for the N-like, O-like, F-like and Ne-like isoelectronic series.

N-like $\|\lambda_\kappa\|$	term	O-like $\|\lambda_\kappa\|$	term	F-like $\|\lambda_\kappa\|$	term	Ne-like $\|\lambda_\kappa\|$	term
2.41491	^{4}S	3.02641	^{3}P	3.68415	^{2}P	4.38541	^{1}S
2.43246	^{2}D	3.03769	^{1}D	3.78926	^{2}S		
2.44111	^{2}P	3.05065	^{1}S				
2.49314	^{4}P	3.11850	^{3}P				
2.52109	^{2}D	3.14982	^{1}P				
2.53864	^{2}S	3.24065	^{1}S				
2.54189	^{2}P						
2.61775	^{2}P						

3.5 General symmetry-adapted basis sets derived from the large-Z approximation

In the general case of an N-electron atom, it would be more complicated to construct symmetry-adapted basis functions of the Russell-Saunders type by means of group theoretical coupling coefficients. Fortunately this is not necessary. We shall see that general symmetry-adapted basis sets can be derived by considering the large-Z approximation.

The $\mathcal{R}$-blocks can in some cases become very large, and in such cases it is desirable to reduce the dimension of the block of $T'_{\nu',\nu}$ that needs to be diagonalized in equation (3.52). This can be done by restricting the primitive Goscinskian configurations in the blocks to those characterized by particular values of M and M_S. The problem that arises if one does this is that the degeneracy of the roots can no longer be used as a clue to the values of the quantum numbers L and S to which the roots and eigenvectors correspond. However, there is a way around this difficulty:

If we diagonalize an entire $\mathcal{R}$-block of the interelectron repulsion matrix we notice that the roots λ_κ are $(2L+1) \times (2S+1)$-fold degenerate. This degeneracy corresponds to the range of values

$$M = L, L-1, L-2, \ldots, -L \qquad (3.69)$$

and

$$M_S = S, S-1, S-2, \ldots, -S \qquad (3.70)$$

Let us now consider the smaller blocks that are characterized by particular positive values of $\mathcal{R}_\nu$, M and M_S. When such a small block is diagonalized, it will contain roots and eigenvectors corresponding to $L \geq M$ and $S \geq M_S$. Within such a small block, the roots λ_κ will be non-degenerate.

Now suppose that we wish to generate a symmetry-adapted basis set corresponding to the quantum numbers L and S. We can do this by first examining the small block characterized by $M = L+1$ and $M_S = S$. We know that all the roots and eigenvectors in this small block will be wrong because they will correspond to values of the total angular momentum quantum number larger than the desired value. Similarly, all of the roots and eigenvectors in the small block with $M = L$ and $M_S = S+1$ will be wrong.

Finally we diagonalize the interelectron repulsion matrix for $M = L$ and $M_S = S$ and discard all of the roots that occurred in the two previously-examined small blocks. The remaining roots and their corresponding eigenvectors will be characterized by the desired values of L and S as well as by the quantum numbers $M = L$ and $M_S = S$. This is because λ_κ values corresponding to larger values of total spin and total angular momentum than the desired values will occur in one or another of the two small blocks previously examined. Roots corresponding to smaller values of total angular momentum and total spin than those desired cannot occur in the small block with $M = L$ and $M_S = S$. The only remaining possibility, when all of the roots of the first two small blocks have been discarded, are roots corresponding to the desired values of L and S.

For particular values of the quantum numbers S and L, the eigenvectors corresponding to $M = L$ and $M_S = S$ will be simpler in form than those corresponding to lower values of M and M_S, and these are exactly the symmetry-adapted basis functions obtained by the method described above. In a case with spherical symmetry, i.e. a case where no external fields are present, it is sufficient to use a basis set characterized by particular values of the azimuthal quantum numbers, since the energies and other properties of an atom or ion are independent of them. The symmetry-adapted basis sets found by this method can be used to perform accurate calculations on the spectra of few-electron atoms.

3.6 Symmetry-adapted basis functions from iteration

The problem of generating and identifying Russell-Saunders basis functions can be solved in another way by using the method of iteration discussed at the end of Chapter 2. Suppose that the ith iterated solution to the Schrödinger equation for an atom or ion is given by

$$\Psi_\kappa^{(i)}(\mathbf{x}) = \sum_{\nu \in I_i} \Phi_\nu(\mathbf{x}) B_{\nu,\kappa}^{(i)} \qquad (3.71)$$

where the basis functions $\Phi_\nu(\mathbf{x})$ are primitive Goscinskian configurations, and suppose that the initial trial function $\Psi_\kappa^{(0)}(\mathbf{x})$ is found by solving the Sturmian secular equations within the domain I_0:

$$\sum_{\nu \in I_0} [T_{\nu',\nu} - p_\kappa \delta_{\nu',\nu}] B_{\nu,\kappa}^{(0)} = 0 \qquad (3.72)$$

Then the coefficients of the first-iterated solution will be given by

$$B_{\nu',\kappa}^{(1)} \sim \sum_{\nu \in I_0} T_{\nu',\nu} B_{\nu,\kappa}^{(0)} \qquad \nu' \in I_1 \qquad (3.73)$$

Primitive configurations with coefficients $B_{\nu',\kappa}^{(1)}$ less than a given magnitude can be discarded. If we set $I_2 = I_1$, the coefficients of the second-iterated solution will be determined by

$$B^{(2)}_{\nu',\kappa} \sim \sum_{\nu \in I_1} T_{\nu',\nu} B^{(1)}_{\nu,\kappa} \qquad \nu' \in I_1 \qquad (3.74)$$

With the proper choice of I_0 and I_1, equations (3.72)-(3.74) can be used to construct Russell-Saunders basis functions of a particular symmetry. To do this, we let I_0 correspond to an $\mathcal{R}$-block that we know contains a solution of the desired symmetry. For example, if we are interested in ^2P states of the lithium-like isoelectronic series, we can begin by diagonalizing $T_{\nu',\nu}$ for the lowest lithium-like $\mathcal{R}$-block. The "target function" $\Psi^{(0)}_\kappa$ can be identified as a ^2P state by its multiplicity. In our example, we might also choose a target function with $M = 1$ and $M_S = \frac{1}{2}$. The parity of this state is odd, since parity is determined by $(-1)^{l_1+l_2+\cdots+l_N}$. The domain I_1 is then chosen to be the union of I_0 and all the Goscinskian configurations that correspond to the same M and M_S values, and the same parity, taken from a variety of other $\mathcal{R}$-blocks. Whenever an $\mathcal{R}$-block is included, *all* of the primitive configurations within the block that meet the M, M_S and parity criteria must be included in I_1 if symmetry is to be maintained, but those corresponding to other values of M, M_S and parity need not be included. We still need to distinguish between the different linearly independent symmetry-adapted basis functions contained in $B^{(2)}_{\nu',\kappa}$. We can do this by carrying out the double iteration from the target function with two different values of the nuclear charge. We then calculate the ratio $B^{(2)}_{\nu',\kappa}(Z_1)/B^{(2)}_{\nu',\kappa}(Z_2)$. This ratio will correspond to a particular number for all coefficients belonging to the same symmetry-adapted basis function, but the ratio will differ for coefficients belonging to different basis functions.

Chapter 4
AUTOIONIZING STATES

4.1 Electron correlation and the molecule-like character of autoionizing states

The doubly excited states of the 2-electron isoelectronic series exhibit a high degree of angular correlation. If we examine the wave functions of such states by requiring the position of one electron to be at a particular position, and if we then compute the probability density for the other electron, we find that it has a well-marked maximum on the opposite side of the nucleus from the first electron. This fact has led a number of authors[1] to compare such doubly-excited states with the rotational and vibrational modes of a triatomic molecule.

Besides exhibiting a high degree of angular correlation, doubly-excited states of the 2-electron isoelectronic series have sufficient energy to ionize spontaneously.

4.2 Calculation of autoionizing states using generalized Sturmians

The generalized Sturmian method, as described in Chapter 3, is very well suited for the rapid calculation of doubly-excited states of atoms and atomic ions. However, in order to accurately represent the very high degree of angular correlation in the wave functions of such states, it is necessary to include orbitals with high values of angular momentum in the Goscinskian configurations used in the calculation.

For the large-Z members of an isoelectronic series, simple approximate expressions for the energies of doubly-excited states can be found by diag-

---

[1]See for example [Macek, 1968], [Lin, 1975], [Klar and Klar, 1980], [Green, 1981], [Read, 1982] and [Berry et al, 1993].

onalizing the interelectron repulsion matrix for the relevant $\mathcal{R}$-block, as is described in Chapter 3, Section 3.4. For example, Table 4.1 shows the roots of the interelectron repulsion matrix $T'_{\nu',\nu}$ for the $n = 2$, $n' = 2$ $\mathcal{R}$-block, and for the $n = 3$, $n' = 3$ $\mathcal{R}$-block. The reader can verify that for the 22-block the sum of the degeneracies is given by the binomial coefficient

$$\binom{8}{2} = 28 \tag{4.1}$$

which is the number of ways in which two different spin-orbitals may be chosen from the eight spin-orbitals of the $n = 2$ shell. For the 33-block, the sum of the degeneracies is the binomial coefficient

$$\binom{18}{2} = 153 \tag{4.2}$$

i.e., the number of ways of choosing two different spin-orbitals from the eighteen available in the $n = 3$ shell. For the 22-block, the energies of the autoionizing states are given approximately by

$$E_\kappa \approx -\frac{1}{2}(Z\mathcal{R} - |\lambda_\kappa|)^2 = -\frac{1}{2}\left(Z\sqrt{\frac{2}{2^2}} - |\lambda_\kappa|\right)^2 \tag{4.3}$$

where the roots $|\lambda_\kappa|$ are to be found in the first column of Table 4.1. The corresponding expression for the 33-block is

$$E_\kappa \approx -\frac{1}{2}(Z\mathcal{R} - |\lambda_\kappa|)^2 = -\frac{1}{2}\left(Z\sqrt{\frac{2}{3^2}} - |\lambda_\kappa|\right)^2 \tag{4.4}$$

where the roots λ_κ are taken from the fourth column of Table 4.1. These approximate expressions become progressively more accurate as the nuclear charge Z increases. As was discussed in Chapter 3, the large-Z approximation consists in limiting the basis set to the Goscinskian configurations belonging to a particular $\mathcal{R}$-block. Thus equation (4.3) results when the basis set is limited to the 28 configurations belonging to the 22-block, while (4.4) results when the basis set is limited to the 153 configurations of the 33-block. A comparison with results obtained using a larger basis set and with experimental results is given in Table 4.2.

Table 4.1: Eigenvalues of the 2-electron interelectron repulsion matrix $T'_{\nu',\nu}$ for $n = 2$, $n' = 2$ and for $n = 3$, $n' = 3$.

22 block $|\lambda_\kappa|$	term	deg.	33 block $|\lambda_\kappa|$	term	deg.
0.173881	^{1}S	1	0.104357	^{1}S	1
0.187825	^{3}P	9	0.107954	^{3}P	9
0.232019	^{3}P	9	0.119441	^{1}D	5
0.261850	^{1}D	5	0.128070	^{3}P	9
0.270689	^{1}P	3	0.133676	^{1}P	3
0.345401	^{1}S	1	0.137739	^{1}D	5
			0.137949	^{3}F	21
			0.145472	^{3}D	15
			0.157994	^{3}D	15
			0.160599	^{1}D	5
			0.165518	^{3}F	21
			0.166683	^{1}S	1
			0.168628	^{3}P	9
			0.190539	^{1}G	9
			0.193443	^{3}P	9
			0.197203	^{1}F	7
			0.222037	^{1}D	5
			0.238660	^{1}P	3
			0.280835	^{1}S	1

Table 4.2: 2-electron autoionizing states of the 22-block. Energies in the large-Z approximation, equation (4.3) (top) are compared with those calculated using a larger basis set (middle) and with experimental values [Vainshtein and Safronova, 1978] (bottom).

term $\|\lambda_\kappa\|$	^{1}S 0.173881	^{3}P 0.187825	^{3}P 0.232019	^{1}D 0.26185	^{1}P 0.270689	^{1}S 0.345401
Z=6	−8.27740 −8.28716 −8.29382	−8.22076 −8.22901 −8.23466	−8.04254 −8.05247 −8.05690	−7.92335 −7.96071 −7.97561	−7.88820 −7.93651 −7.95097	−7.59424 −7.65043 −7.65478
Z=7	−11.4045 −11.4144 −11.4247	−11.3380 −11.3462 −11.3551	−11.1285 −11.1384 −11.1449	−10.9882 −11.0258 −11.0421	−10.9468 −10.9958 −11.0125	−10.6000 −10.6565 −10.6625
Z=8	−15.0315 −15.0416 −15.0579	−14.9551 −14.9635 −14.9787	−14.7144 −14.7244 −14.7345	−14.5530 −14.5907 −14.6109	−14.5054 −14.5549 −14.5759	−14.1058 −14.1625 −14.1726
Z=9	−19.1585 −19.1687 −19.1908	−19.0723 −19.0807 −19.1019	−18.8004 −18.8103 −18.8224	−18.6179 −18.6557 −18.6765	−18.5640 −18.6139 −18.6398	−18.1115 −18.1684 −18.1809
Z=10	−23.7856 −23.7959 −23.8309	−23.6895 −23.6979 −23.7308	−23.3863 −23.3963 −23.4160	−23.1827 −23.2206 −23.2493	−23.1226 −23.1728 −23.2096	−22.6173 −22.6743 −22.6945
Z=11	−28.9126 −28.9230 −28.9737	−28.8067 −28.8151 −28.8632	−28.4722 −28.4822 −28.5122	−28.2476 −28.2855 −28.3226	−28.1812 −28.2317 −28.2800	−27.6231 −27.6802 −27.7115
Z=12	−34.5397 −34.5501 −34.6229	−34.4239 −34.4323 −34.4999	−34.0582 −34.0681 −34.1094	−33.8124 −33.8504 −33.8974	−33.7398 −33.7905 −33.8553	−33.1288 −33.1860 −33.2293
Z=13	−40.6667 −40.6772 −40.7774	−40.5411 −40.5495 −40.6428	−40.1441 −40.1541 −40.2111	−39.8773 −39.9153 −39.9757	−39.7984 −39.8493 −39.9357	−39.1346 −39.1919 −39.2518
Z=14	−47.2938 −47.3043 −47.4389	−47.1583 −47.1667 −47.2924	−46.7300 −46.7400 −46.8168	−46.4421 −46.4802 −46.5572	−46.3569 −46.4080 −46.5216	−45.6404 −45.6977 −45.7780
Z=15	−54.4208 −54.4314 −54.6102	−54.2755 −54.2839 −54.4500	−53.8160 −53.8260 −53.9281	−53.5069 −53.5451 −53.6423	−53.4155 −53.4667 −53.6145	−52.6461 −52.7035 −52.8086
Z=16	−62.0479 −62.0585 −62.2912	−61.8926 −61.9011 −62.1168	−61.4019 −61.4119 −61.5458	−61.0718 −61.1100 −61.2304	−60.9741 −61.0254 −61.2143	−60.1519 −60.2094 −60.3445

4.3 Higher series of ^{3}S autoionizing states

In Chapter 3, Tables 3.4 and 3.5, and Figure 3.1, we showed a series of ^{3}S autoionizing states with $n = 2$ and $n' = 3, 4, 5, \ldots$ (see also Figure 4.1). Table 4.3 and Figure 4.2 show a second series of doubly-excited ^{3}S states of the 2-electron isoelectronic series. For this second series, $n = 3$ and $n' = 4, 5, 6, 7, \ldots$. Figure 4.2 shows the ratio E_κ/Z^2 for these states as a function of Z. A third series of doubly-excited ^{3}S states is shown in Table 4.4 and Figure 4.3. For the third series, $n = 4$ and $n' \geq 5$. A fourth series of doubly-excited ^{3}S states is given in Table 4.5. We believe that the calculated energies are accurate, but unfortunately there are no experimental values available with which they can be compared.

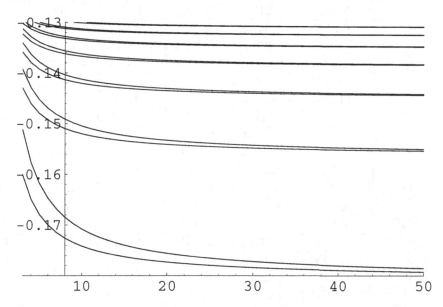

Fig. 4.1 E_κ/Z^2 as a function of Z for the first series of doubly-excited ^{3}S states of the 2-electron isoelectronic series (Table 4.1). For this series, $n=2$, while $n' = 3, 4, 5, 6, \ldots$. In the large-Z limit, E_κ/Z^2 approaches the Z-independent value, $-\mathcal{R}_\nu^2/2$.

Table 4.3: Autoionizing ^{3}S states of heliumlike atoms with $n = 3$ and $n' = 4, 5$. See Figure 4.2. No experimental values of the energies are available.

	3s4s ^{3}S	3p4p ^{3}S	3d4d ^{3}S	3s5s ^{3}S	3p5p ^{3}S	3d5d ^{3}S
C^{4+}	−2.9372	−2.8729	−2.7660	−2.5788	−2.5402	−2.4867
N^{5+}	−4.0338	−3.9576	−3.8293	−3.5368	−3.4908	−3.4260
O^{6+}	−5.3040	−5.2159	−5.0662	−4.6459	−4.5924	−4.5164
F^{7+}	−6.7477	−6.6478	−6.4768	−5.9061	−5.8452	−5.7579
Ne^{8+}	−8.3651	−8.2534	−8.0609	−7.3175	−7.2491	−7.1505
Na^{9+}	−10.1561	−10.0325	−9.8187	−8.8799	−8.8041	−8.6942
Mg^{10+}	−12.1207	−11.9852	−11.7501	−10.5934	−10.5101	−10.3890
Al^{11+}	−14.2590	−14.1116	−13.8551	−12.4581	−12.3673	−12.2349
Si^{12+}	−16.5708	−16.4116	−16.1337	−14.4739	−14.3757	−14.2319
P^{13+}	−19.0562	−18.8851	−18.5860	−16.6408	−16.5351	−16.3801
S^{14+}	−21.7153	−21.5323	−21.2118	−18.9588	−18.8456	−18.6793
Cl^{15+}	−24.5480	−24.3531	−24.0113	−21.4279	−21.3073	−21.1297
Ar^{16+}	−27.5542	−27.3475	−26.9844	−24.0481	−23.9200	−23.7312
K^{17+}	−30.7341	−30.5155	−30.1310	−26.8194	−26.6839	−26.4838
Ca^{18+}	−34.0876	−33.8572	−33.4513	−29.7419	−29.5989	−29.3875
Sc^{19+}	−37.6147	−37.3724	−36.9452	−32.8154	−32.6650	−32.4423
Ti^{20+}	−41.3154	−41.0612	−40.6128	−36.0401	−35.8822	−35.6482
V^{21+}	−45.1898	−44.9237	−44.4539	−39.4159	−39.2505	−39.0053
Cr^{22+}	−49.2377	−48.9598	−48.4686	−42.9428	−42.7699	−42.5134
Mn^{23+}	−53.4593	−53.1695	−52.6570	−46.6208	−46.4404	−46.1727
Fe^{24+}	−57.8544	−57.5527	−57.0189	−50.4499	−50.2621	−49.9831
Co^{25+}	−62.4232	−62.1096	−61.5545	−54.4301	−54.2348	−53.9446
Ni^{26+}	−67.1656	−66.8402	−66.2637	−58.5614	−58.3587	−58.0572
Cu^{27+}	−72.0816	−71.7443	−71.1465	−62.8439	−62.6337	−62.3209
Zn^{28+}	−77.1712	−76.8220	−76.2029	−67.2774	−67.0598	−66.7357

Table 4.4: The third series of heliumlike autoionizing states. Here $n=4$, while $n'=5, 6$. These states correspond to the lowest curves in Figure 4.3. Energies are given in Hartrees. No experimental values are available.

	4s5s ^{3}S	4p5p ^{3}S	4d5d ^{3}S	4f5f ^{3}S	4s6s ^{3}S	4p6p ^{3}S
C^{4+}	−1.7339	−1.7058	−1.6637	−1.5960	−1.5350	−1.5153
N^{5+}	−2.3813	−2.3482	−2.2980	−2.2163	−2.1064	−2.0830
O^{6+}	−3.1312	−3.0930	−3.0349	−2.9391	−2.7682	−2.7410
F^{7+}	−3.9836	−3.9404	−3.8743	−3.7644	−3.5202	−3.4893
Ne^{8+}	−4.9386	−4.8902	−4.8161	−4.6922	−4.3625	−4.3279
Na^{9+}	−5.9960	−5.9426	−5.8605	−5.7226	−5.2950	−5.2568
Mg^{10+}	−7.1559	−7.0974	−7.0073	−6.8554	−6.3179	−6.2759
Al^{11+}	−8.4183	−8.3548	−8.2566	−8.0907	−7.4310	−7.3853
Si^{12+}	−9.7832	−9.7146	−9.6085	−9.4286	−8.6344	−8.5850
P^{13+}	−11.2507	−11.1770	−11.0628	−10.8689	−9.9281	−9.8750
S^{14+}	−12.8206	−12.7418	−12.6197	−12.4118	−11.3120	−11.2552
Cl^{15+}	−14.4930	−14.4092	−14.2790	−14.0571	−12.7863	−12.7258
Ar^{16+}	−16.2679	−16.1790	−16.0408	−15.8050	−14.3508	−14.2866
K^{17+}	−18.1453	−18.0513	−17.9051	−17.6554	−16.0056	−15.9377
Ca^{18+}	−20.1252	−20.0262	−19.8720	−19.6082	−17.7506	−17.6790
Sc^{19+}	−22.2077	−22.1035	−21.9413	−21.6636	−19.5860	−19.5107
Ti^{20+}	−24.3926	−24.2834	−24.1131	−23.8214	−21.5116	−21.4326
V^{21+}	−26.6800	−26.5657	−26.3875	−26.0818	−23.5275	−23.4448
Cr^{22+}	−29.0699	−28.9506	−28.7643	−28.4447	−25.6337	−25.5472
Mn^{23+}	−31.5623	−31.4379	−31.2436	−30.9100	−27.8301	−27.7400
Fe^{24+}	−34.1573	−34.0278	−33.8254	−33.4779	−30.1169	−30.0230
Co^{25+}	−36.8547	−36.7201	−36.5098	−36.1483	−32.4939	−32.3963
Ni^{26+}	−39.6546	−39.5149	−39.2966	−38.9211	−34.9612	−34.8599
Cu^{27+}	−42.5570	−42.4123	−42.1859	−41.7965	−37.5187	−37.4137
Zn^{28+}	−45.5619	−45.4121	−45.1777	−44.7744	−40.1666	−40.0579

Table 4.5: The fourth series of heliumlike ^{3}S autoionizing states.

	5s6s ^{3}S	5p6p ^{3}S	5d6d ^{3}S	5f6f ^{3}S	5g6g ^{3}S	5s7s ^{3}S
C^{4+}	−1.1467	−1.1322	−1.1116	−1.0806	−1.0341	−1.0252
N^{5+}	−1.5749	−1.5577	−1.5334	−1.4967	−1.4401	−1.4073
O^{6+}	−2.0708	−2.0511	−2.0230	−1.9806	−1.9138	−1.8498
F^{7+}	−2.6345	−2.6122	−2.5803	−2.5322	−2.4554	−2.3527
Ne^{8+}	−3.2660	−3.2411	−3.2055	−3.1517	−3.0648	−2.9160
Na^{9+}	−3.9653	−3.9378	−3.8984	−3.8389	−3.7420	−3.5397
Mg^{10+}	−4.7323	−4.7022	−4.6591	−4.5939	−4.4870	−4.2238
Al^{11+}	−5.5672	−5.5344	−5.4875	−5.4166	−5.2998	−4.9684
Si^{12+}	−6.4698	−6.4344	−6.3838	−6.3071	−6.1804	−5.7733
P^{13+}	−7.4402	−7.4022	−7.3478	−7.2655	−7.1288	−6.6387
S^{14+}	−8.4783	−8.4378	−8.3796	−8.2916	−8.1450	−7.5644
Cl^{15+}	−9.5842	−9.5411	−9.4792	−9.3854	−9.2289	−8.5506
Ar^{16+}	−10.7580	−10.7122	−10.6465	−10.5471	−10.3807	−9.5972
K^{17+}	−11.9995	−11.9511	−11.8817	−11.7765	−11.6002	−10.7042
Ca^{18+}	−13.3087	−13.2578	−13.1846	−13.0737	−12.8875	−11.8715
Sc^{19+}	−14.6858	−14.6323	−14.5553	−14.4387	−14.2426	−13.0993
Ti^{20+}	−16.1306	−16.0745	−15.9937	−15.8715	−15.6654	−14.3876
V^{21+}	−17.6432	−17.5845	−17.5000	−17.3720	−17.1561	−15.7362
Cr^{22+}	−19.2236	−19.1623	−19.0740	−18.9403	−18.7145	−17.1452
Mn^{23+}	−20.8717	−20.8078	−20.7158	−20.5764	−20.3407	−18.6146
Fe^{24+}	−22.5877	−22.5212	−22.4254	−22.2803	−22.0347	−20.1445
Co^{25+}	−24.3714	−24.3023	−24.2027	−24.0519	−23.7965	−21.7347
Ni^{26+}	−26.2229	−26.1512	−26.0479	−25.8914	−25.6260	−23.3854
Cu^{27+}	−28.1421	−28.0678	−27.9608	−27.7986	−27.5233	−25.0965
Zn^{28+}	−30.1292	−30.0523	−29.9415	−29.7735	−29.4884	−26.8679

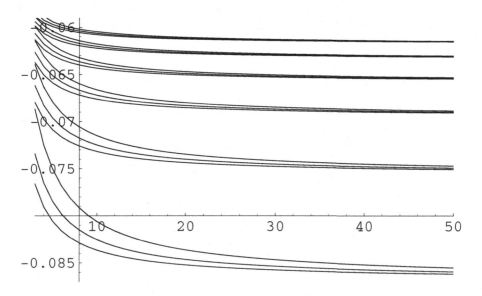

Fig. 4.2 E_κ/Z^2 as a function of Z for the second series of doubly-excited ^{3}S states of heliumlike atoms and ions (Table 4.2). For this series, $n=3$, while $n' = 4, 5, 6, 7, \ldots$.

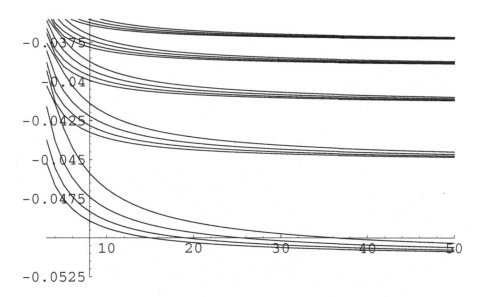

Fig. 4.3 This figure is similar to Figures 4.1 and 4.2, but it shows the third series of autoionizing ^{3}S states (Table 4.3). For this series $n=4$, while $n' = 5, 6, 7, 8, \ldots$.

Chapter 5

CORE IONIZATION

5.1 Core ionization energies in the large-Z approximation

As a further illustration of the general principles discussed in Chapter 3 we have calculated the core ionization energies of a number of atoms and ions in the large-Z approximation. From equation (3.55) we can see that, in this rough approximation, the energy of the ground state or an excited state of an atom or ion can be expressed as a quadratic function of the nuclear charge, Z. For example, the ground-state energies of the first few N-electron isoelectronic series are given approximately by

$$E_0 \approx -\tfrac{1}{2}\left(Z\sqrt{\tfrac{2}{1}} - 0.441942\right)^2, \quad N = 2$$

$$E_0 \approx -\tfrac{1}{2}\left(Z\sqrt{\tfrac{2}{1}+\tfrac{1}{4}} - 0.681870\right)^2, \, N = 3$$

$$E_0 \approx -\tfrac{1}{2}\left(Z\sqrt{\tfrac{2}{1}+\tfrac{2}{4}} - 0.986172\right)^2, \, N = 4$$

$$E_0 \approx -\tfrac{1}{2}\left(Z\sqrt{\tfrac{2}{1}+\tfrac{3}{4}} - 1.40355\right)^2, \, N = 5 \qquad (5.1)$$

$$E_0 \approx -\tfrac{1}{2}\left(Z\sqrt{\tfrac{2}{1}+\tfrac{4}{4}} - 1.88151\right)^2, \, N = 6$$

$$E_0 \approx -\tfrac{1}{2}\left(Z\sqrt{\tfrac{2}{1}+\tfrac{5}{4}} - 2.41491\right)^2, \, N = 7$$

where the roots of the interelectron repulsion matrices for the $\mathcal{R}$-blocks have been taken from Tables 3.6, 3.7 and 3.8. This is a rough approximation, but it yields energies in reasonably good agreement with Clementi's Hartree-Fock results [Clementi, 1963], as is shown in our previous publications. We now remove an electron from the 1s orbital of the core of each atom or ion, so that (for example)

$$|\chi_{2s}\chi_{1s}\chi_{\bar{1}s}| \rightarrow |\chi_{2s}\chi_{1s}| \qquad N = 3 \qquad (5.2)$$

We thus obtain a set of core-excited states for which we can calculate the approximate energy in the large-Z approximation, using equation (3.55). This gives us the energies

$$E_i \approx -\tfrac{1}{2}\left(Z\sqrt{\tfrac{1}{1}}\right)^2, \qquad N' = 1$$

$$E_i \approx -\tfrac{1}{2}\left(Z\sqrt{\tfrac{1}{1}+\tfrac{1}{4}} - 0.168089\right)^2, N' = 2$$

$$E_i \approx -\tfrac{1}{2}\left(Z\sqrt{\tfrac{1}{1}+\tfrac{2}{4}} - 0.433936\right)^2, N' = 3$$

$$E_i \approx -\tfrac{1}{2}\left(Z\sqrt{\tfrac{1}{1}+\tfrac{3}{4}} - 0.800757\right)^2, N' = 4 \qquad (5.3)$$

$$E_i \approx -\tfrac{1}{2}\left(Z\sqrt{\tfrac{1}{1}+\tfrac{4}{4}} - 1.23703\right)^2, N' = 5$$

$$E_i \approx -\tfrac{1}{2}\left(Z\sqrt{\tfrac{1}{1}+\tfrac{5}{4}} - 1.73489\right)^2, N' = 6$$

In equation (5.3), $N' = N - 1$ represents the number of electrons after core ionization. From among the possible states, the core-ionized configuration has been taken to be that of minimum energy. In other words, the smallest root $|\lambda_\kappa|$ of the interelectron repulsion matrix has been chosen from Tables 5.1 and 5.2 The energies of other core-ionized configurations could be calculated by choosing other roots. By subtracting (5.3) from (5.1), we obtain the approximate core ionization energies in Hartrees as quadratic functions of the nuclear charge, Z:

$$\Delta E \approx 0.097656 - 0.625000Z + 0.5Z^2, \ N = 2$$

$$\Delta E \approx 0.218346 - 0.834876Z + 0.5Z^2, \ N = 3$$

$$\Delta E \approx 0.393436 - 1.02781Z + 0.5Z^2, \ N = 4$$

$$\Delta E \approx 0.664370 - 1.26822Z + 0.5Z^2, \ N = 5 \quad (5.4)$$

$$\Delta E \approx 1.00492 - 1.50945Z + 0.5Z^2, \ N = 6$$

$$\Delta E \approx 1.41098 - 1.75121Z + 0.5Z^2, \ N = 7$$

The approximate ionization energies of equation (5.4) are shown in Table 5.3 and in Figures 5.1 and 5.2.

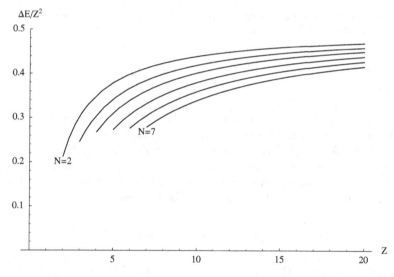

Fig. 5.1 The energy ΔE required to remove an electron from the inner shell of an N-electron atom or ion is an approximately quadratic function of the nuclear charge Z (equation (5.4)).

GENERALIZED STURMIANS AND ATOMIC SPECTRA

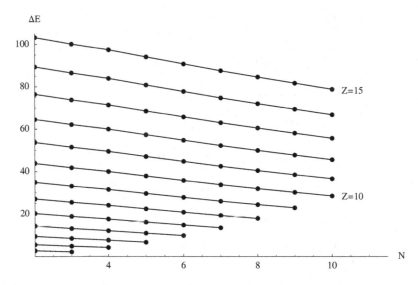

Fig. 5.2 The core ionization energy ΔE plotted as a function of N for several values of Z. The approximately linear dependence on N is discussed in the text.

5.2 Isonuclear series; piecewise-linear dependence of ΔE on N

The approximately linear dependence of the core ionization energy ΔE on the number of electrons N (Figure 5.2) can be understood by means of the following crude model: Let $2 \leq N \leq 10$. Suppose that the $N-2$ electrons in the $n=2$ shell have a sharply peaked density distribution $\rho(r)$ with a maximum at $r = a$. In a very rough approximation, we can imagine that the charge density is localized on a thin shell so that

$$\rho(r) = \frac{N-2}{4\pi a^2}\delta(r-a) \tag{5.5}$$

The potential due to these electrons will then be given by

$$v(r) = \int d^3x' \frac{\rho(r')}{|\mathbf{x}-\mathbf{x}'|} = \frac{N-2}{a^2}\int_0^\infty dr' \frac{r'^2}{r_>}\delta(r'-a) \tag{5.6}$$

$$= \begin{cases} \dfrac{N-2}{r} & \text{for } r > a \\[6pt] \dfrac{N-2}{a} & \text{for } r \leq a \end{cases} \tag{5.7}$$

In other words, the potential in the core region is linear in N. The core ionization energy ΔE has two components: Firstly there is the effect of the nucleus and the remaining core electron (and this is independent of N). Secondly, there is the term $\frac{N-2}{a}$, which is due to the electrons in the $n = 2$ shell, and this second term must be subtracted from the first when we calculate ΔE. Of course, the $N - 2$ electrons with $n = 2$ have a density $\rho(r)$ that is not really so sharply peaked, but nevertheless one can obtain a qualitative understanding from the crude model. For $N > 10$, the dependence on N is approximately piecewise linear, with changes in slope occurring when each new shell begins to fill, as is illustrated in Figure 5.3.

Having used a very crude model to obtain a qualitative understanding of the approximately linear dependence of ΔE on N, it is interesting to ask how the radial density distribution actually looks. Figure 5.4 shows the ground state radial density distribution $\rho(r)r^2$ for $Z = 18$ and $1 \leq N \leq 18$. The various curves in the figure correspond to various values of N. The contribution of the $n = 2$ shell is shown in Figure 5.5 for $3 \leq N \leq 10$. The curves in the figure correspond to various values of N. If we let $\rho(r)$ be the spherically averaged density due to the electrons in the $n = 2$ shell, then the electrostatic potential produced at the position of the nucleus by these electrons will be

$$v(0) = \int d^3x' \, \frac{\rho(\mathbf{x}')}{|\mathbf{x}'|} = 4\pi \int_0^\infty dr' \, r'\rho(r') \qquad (5.8)$$

Figure 5.6 shows $-v(0)$ as a function of N for $Z = 18$ and $2 < N \leq 10$. We can see in the figure the approximately linear dependence on N that is also apparent in ΔE (Figure 5.2).

5.3 Core ionization energies for the 3-electron isoelectronic series

Although the large-Z approximation can give us a rough general view of the behavior of core ionization energies as functions of N and Z, the accuracy of the approximation is rather low. Therefore it is interesting to try to make a better calculation using a larger number of Goscinskian configurations. For example, calculated core-ionization energies for the 3-electron isoelectronic series are shown Table 5.4. The calculated energies of the ^3S core-ionized states of the series are compared in the table with experimental energies from the NIST Tables. Symmetry-adapted configurations used in the calculation were generated using the iteration method discussed in Chapter 2.

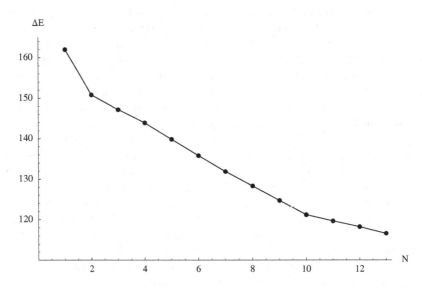

Fig. 5.3 The core ionization energy ΔE plotted as a function of N for $Z = 18$ with $1 \leq N \leq 13$. The N-dependence is approximately piecewise linear, with a breaks in slope occurring when the $n = 2$ and the $n = 3$ shells begin to fill.

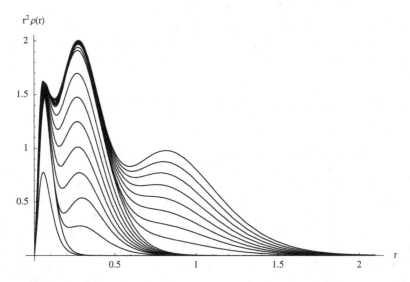

Fig. 5.4 The total radial density distribution $\rho(r)r^2$ for the ground state of the isonuclear series with $Z = 18$. Each curve corresponds to a value of N, with $1 \leq N \leq 18$.

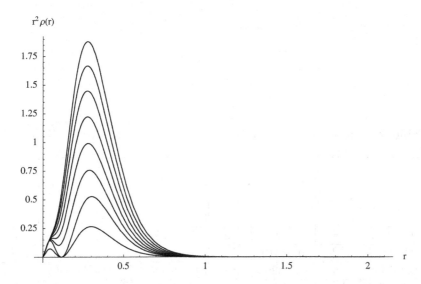

Fig. 5.5 The contribution to the radial density distribution from electrons in $n = 2$ shell. The curves correspond to various values of N, with $2 < N \leq 10$. As in the previous figure, $Z = 18$.

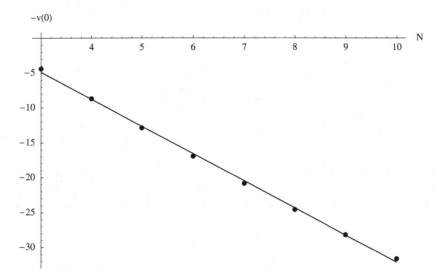

Fig. 5.6 The electrostatic potential $-v(0)$ at the nucleus due to the electrons in the $n = 2$ shell for $Z = 18$ (equation (5.8)). The radial density distributions shown in figure 5.3 were used in calculating $v(0)$. The dependence on N is approximately linear.

Table 5.1: Roots of the interelectron repulsion matrix for $\mathcal{R}_\nu = \sqrt{\dfrac{1}{1^2} + \dfrac{N'-1}{2^2}}$.

$N'=2$ $\|\lambda_\kappa\|$	term	$N'=3$ $\|\lambda_\kappa\|$	term	$N'=4$ $\|\lambda_\kappa\|$	term	$N'=5$ $\|\lambda_\kappa\|$	term
0.168089	^{3}S	0.433936	^{2}S	0.800757	^{3}P	1.23703	^{4}P
0.201897	^{3}P	0.44619	^{4}P	0.808142	^{5}P	1.24149	^{6}S
0.207350	^{1}S	0.493265	^{2}P	0.826565	^{1}P	1.26402	^{2}D
0.232434	^{1}P	0.502568	^{4}P	0.859138	^{3}D	1.27324	^{2}P
		0.510673	^{2}P	0.864205	^{3}P	1.27824	^{2}S
		0.533729	^{2}D	0.883056	^{3}S	1.29665	^{4}D
		0.544383	^{2}P	0.883961	^{5}S	1.30582	^{4}S
		0.573228	^{2}S	0.890413	^{3}P	1.31157	^{4}P
				0.892320	^{1}D	1.33742	^{2}D
				0.913294	^{1}P	1.33935	^{4}S
				0.916238	^{1}S	1.35234	^{2}P
				0.920783	^{3}D	1.35250	^{2}D
				0.935578	^{3}S	1.36741	^{2}P
				0.943943	^{3}P	1.37264	^{2}S
				0.946592	^{1}D	1.38679	^{4}P
				0.969751	^{1}P	1.41378	^{2}D
						1.42301	^{2}P
						1.44431	^{2}S

Table 5.2: Roots of the interelectron repulsion matrix for $\mathcal{R}_\nu = \sqrt{\frac{1}{1^2} + \frac{N'-1}{2^2}}$.

$N'=6$ $\|\lambda_\kappa\|$	term	$N'=7$ $\|\lambda_\kappa\|$	term	$N'=8$ $\|\lambda_\kappa\|$	term	$N'=9$ $\|\lambda_\kappa\|$	term
1.73489	^{5}S	2.33058	^{4}P	2.97391	^{3}P	3.66181	^{2}S
1.76736	^{3}D	2.35472	^{2}D	2.9945	^{1}P		
1.77673	^{3}P	2.36297	^{2}P	3.07386	^{3}S		
1.78041	^{3}S	2.36890	^{2}S	3.10033	^{1}S		
1.79012	^{1}D	2.41564	^{4}P				
1.79949	^{1}P	2.45210	^{2}P				
1.80371	^{5}P	2.46559	^{2}P				
1.84869	^{3}D	2.55129	^{2}S				
1.85315	^{3}P						
1.86978	^{3}S						
1.87627	^{3}P						
1.87795	^{1}D						
1.89645	^{1}P						
1.89904	^{1}S						
1.94394	^{3}P						
1.96670	^{1}P						

Table 5.3: Core ionization energies in Hartrees for a number of atoms and ions, calculated in the crude large-Z approximation. In this table, the core-ionized state is always chosen to be the one that has the lowest energy.

	$N=2$	$N=3$	$N=4$	$N=5$	$N=6$
$Z=2$	0.8477				
$Z=3$	2.723	2.214			
$Z=4$	5.598	4.879	4.281		
$Z=5$	9.473	8.544	7.753	6.823	
$Z=6$	14.35	13.21	12.23	11.06	9.948
$Z=7$	20.22	18.87	17.70	16.29	14.94
$Z=8$	27.10	25.54	24.17	22.52	20.93
$Z=9$	34.97	33.20	31.64	29.75	27.92
$Z=10$	43.85	41.87	40.17	38.01	35.91

Table 5.4: Core ionization energies in Hartrees for the 3-electron isoelectronic series. E_0 is the ground-state energy, while E_i is the energy of the ^3S state obtained by removing an electron from the core of the atom or ion. The core-ionization energies, $\Delta E_{calc} = (E_i)_{calc} - (E_0)_{calc}$, are compared with experimental values constructed from the NIST tables [NIST database]. Discrepancies are mainly due to inaccuracy of the calculated ground states.

	$(E_0)_{calc}$	$(E_0)_{expt}$	$(E_i)_{calc}$	$(E_i)_{expt}$	ΔE_{calc}	ΔE_{expt}
Li	-7.45502	-7.47798	-5.10868	-5.11086	2.34634	2.36712
Be^+	-14.2949	-14.3258	-9.29450	-9.29838	5.00036	5.02739
B^{2+}	-23.3897	-23.4289	-14.7308	-14.7378	8.65882	8.69110
C^{3+}	-34.7371	-34.7860	-21.4175	-21.4293	13.3196	13.3567
N^{4+}	-48.336	-48.3991	-29.3542	-29.3745	18.9818	19.0246
O^{5+}	-64.1859	-64.2680	-38.5410	-38.5765	25.6448	25.6916
F^{6+}	-82.2863	-82.3997	-48.9779	-49.0306	33.3083	33.3691
Ne^{7+}	-102.637	-102.792	-60.6649	-60.7446	41.9722	42.0477

Chapter 6

STRONG EXTERNAL FIELDS

6.1 External electric fields

In this chapter, we shall consider the effect of strong external fields on the spectra of atoms and ions. A uniform electric field in the z-direction can be represented by a potential of the form

$$V''(\mathbf{x}) = \sum_{j=1}^{N} \mathcal{E} r_j \cos\theta_j = \mathcal{E} \sum_{j=1}^{N} z_j \tag{6.1}$$

If we introduce the definition

$$T^{el.}_{\nu',\nu} \equiv -p_\kappa \int d x\, \Phi^*_{\nu'}(\mathbf{x}) \sum_{j=1}^{N} z_j\, \Phi_\nu(\mathbf{x}) \tag{6.2}$$

then the Sturmian secular equations, modified by the addition of the electric field, become

$$\sum_{\nu} \left[\delta_{\nu',\nu} Z\mathcal{R} + T'_{\nu',\nu} + \eta T^{el.}_{\nu',\nu} - p_\kappa \delta_{\nu',\nu} \right] B_{\nu,\kappa} = 0 \tag{6.3}$$

where the parameter η is defined by

$$\eta \equiv \frac{\mathcal{E}}{p_\kappa^2} \tag{6.4}$$

Here $\mathcal{E}$ is the electric field strength expressed in atomic units, i.e., Hartrees/electron-Bohr[1]. In order to find the atomic or ionic spectrum as a function of electric field strength, we begin by evaluating the matrix $T^{el.}_{\nu',\nu}$. This matrix is independent of p_κ, and therefore it can be evaluated once and for all, as can $T'_{\nu',\nu}$. With these two matrices in hand, the

---
[1] 1 Hartree/electron-Bohr= 5.142206×10^9 Volts/cm.

Sturmian secular equations (6.3) can then be solved for many values of the parameter η. This gives us a set of curves $p_\kappa(\eta)$ for the various states

$$\Psi_\kappa = \sum_\nu \Phi_\nu B_{\nu,\kappa} \qquad (6.5)$$

Then with the help of (6.4) and

$$E_\kappa(\eta) = -\frac{p_\kappa^2(\eta)}{2} \qquad (6.6)$$

we can obtain a set of curves $E_\kappa(\mathcal{E})$, i.e., curves representing the energies of the ground state and the excited states of the atom or ion as functions of the applied field strength. For a given fixed number of electrons N, the process may be repeated for many values of the nuclear charge Z without re-evaluating the matrices $T'_{\nu',\nu}$ and $T^{el.}_{\nu',\nu}$. All the members of an isoelectronic series can be studied for all values of applied field without re-evaluating these two matrices. Thus the generalized Sturmian method is very convenient and rapid for Stark effect studies.

6.2 Anomalous states

Figure 6.1 shows the energies of triplet excited states of helium as functions of the external applied field $\mathcal{E}$. In addition to the expected Stark splitting of the levels, a number of anomalous states also appear in the figure. As the external field increases in magnitude, the energies of the anomalous states fall rapidly and almost linearly.

The appearance of these anomalous states in the energy spectrum of an atom or ion can be interpreted as follows: When a uniform time-independent external field is applied to an atom, the potential far away from the atom resembles a hill of uniform slope. Near to the atom, there is a local depression in the hill, in which the electrons are temporarily trapped. Over a long period of time, the electrons would tunnel out of this trap into the regions of very low potential lying downhill from the atom.

In our method of treating the Stark effect, we are studying the time-independent Schrödinger equation of this hill-like system. If our basis set is sufficiently rich, we will obtain some solutions that lie primarily inside the atom, and others that lie primarily outside the local trap and downhill from it, as is illustrated in Figure 6.2. The solutions most of whose probability density is outside the atom or ion correspond to the anomalous states, and of course the energy of these states depends strongly and almost linearly on the value of the external electric field. As we add more and more diffuse functions to our basis set, the number of anomalous states in our set of

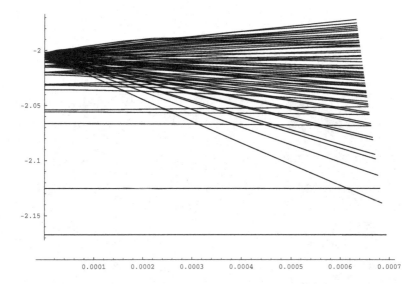

Fig. 6.1 Triplet excited states of helium in a strong external electric field. Both the energies and the field strengths are in atomic units. Only the $M = 0$ states are shown in the figure. A generalized Sturmian basis set with 80 basis functions was used. Besides the expected Stark splitting of the excited states, one can observe in the spectrum a number of anomalous states, whose energy decreases rapidly as the field strength increases. The interpretation of the anomalous states is discussed in the text.

calculated solutions increases. In fact, these states are members of a continuum, and with a basis set of the Goscinskian type, we cannot represent them properly. No matter how many basis functions of the Goscinskian type one adds, the anomalous Stark effect states are always strongly basis-set-dependent. This observation agrees with the remark that we made in the Preface: Generalized Sturmian basis sets, as defined in this book, are only appropriate for the representation of bound states. Since the anomalous states are not bound, they cannot be adequately represented using a generalized Sturmian basis set.

Figure 6.2 shows the excited-state 1-electron orbitals for the first few triplet excited states of helium in a uniform external electric field of 4.5×10^{-5} Hartree/electron-Bohr. Some of the states are normal, while others are anomalous. In making this figure, we have assumed that the orbital for the inner electron is approximately the same for all the configurations that enter the wave function, so that it can be factored out.

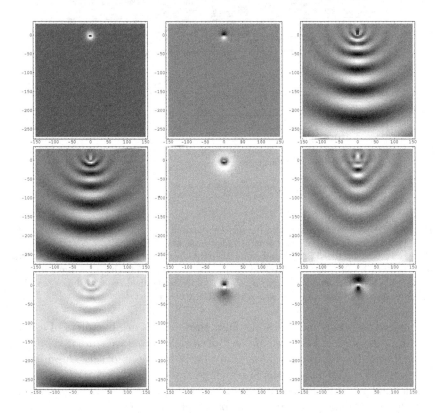

Fig. 6.2 This figure shows cross-sections of the wave function for the first 9 states of a He atom experiencing a field strength of $4.5 \cdot 10^{-5}$ Hartrees/electron-Bohr. The scale of the figures is such that the domain of the atom occupies only a small part of the figure, as can be seen from the normal states (state 1,2,5,8 and 9). Thus the orbitals of the anomalous states are those of electrons that are outside of the atom, only a small part of the wave functions being inside it. With a larger basis set, the orbital part would extend still further from the atom. In other words the orbitals are those of ionized electrons in the field of the nucleus and the remaining bound electron.

Table 6.1: Calculated static polarizabilities of the ground state (in atomic units) for the first few isoelectronic series. N is the number of electrons in the atom or ion, while Z is the nuclear charge.

	$N = 1$	$N = 2$	$N = 3$	$N = 4$
$Z = 1$	4.50000			
$Z = 2$	0.28125	1.50824		
$Z = 3$	0.05557	0.19281	193.247	
$Z = 4$	0.01758	0.05035	26.8171	39.0524
$Z = 5$	0.00720	0.01849	8.28658	10.9107
$Z = 6$	0.00347	0.00831	3.56008	4.59854
$Z = 7$	0.00187	0.00427	1.84243	2.36406
$Z = 8$	0.00110	0.00241	1.07471	1.37379
$Z = 9$	0.00069	0.00146	0.68124	0.86829
$Z = 10$	0.00045	0.00094	0.45896	0.58401

Table 6.2: Approximate induced transition dipole moments between the ground state and some excited ^{1}S and ^{1}D states of C^{2+} for various values of applied electric field. Atomic units are used throughout.

	0.0000	0.0001	0.0002	0.0003	0.0004
$^1S, 1s^22s4s$	0.00000	0.000659	0.001319	0.001978	0.002637
$^1D, 1s^22s4d$	0.00000	0.000957	0.001913	0.002869	0.003825
$^1S, 1s^22s5s$	0.00000	0.002155	0.004306	0.006450	0.008581
$^1D, 1s^22s5d$	0.00000	0.002794	0.005420	0.007825	0.010038
$^1S, 1s^22s6s$	0.00000	0.004878	0.009662	0.014264	0.018614
$^1D, 1s^22s6d$	0.00000	0.001546	0.005251	0.009730	0.014691
$^1S, 1s^22s7s$	0.00000	0.002612	0.006713	0.010685	0.014439

6.3 Polarizabilities

From equations (6.2) and (6.5) it follows that

$$\int dx\ \Psi^*_\kappa(\mathbf{x}) \sum_{j=1}^{N} z_j \Psi_\kappa(\mathbf{x}) = -\frac{1}{p_\kappa} \sum_{\nu,\nu'} B^*_{\nu,\kappa} T^{el.}_{\nu,\nu'} B_{\nu',\kappa} \qquad (6.7)$$

If we also make use of (6.4), we have

$$\frac{1}{\mathcal{E}} \int dx\ \Psi^*_\kappa(\mathbf{x}) \sum_{j=1}^{N} z_j \Psi_\kappa(\mathbf{x}) = -\frac{1}{\eta p_\kappa^3} \sum_{\nu,\nu'} B^*_{\nu,\kappa} T^{el.}_{\nu,\nu'} B_{\nu',\kappa} \qquad (6.8)$$

The expression on the left-hand side of (6.8) is the dipole moment of the state Ψ_κ divided by the external electric field strength $\mathcal{E}$, both expressed in atomic units. Thus equation (6.8) allows us to calculate the polarizability of any state of an atom or ion. We do so by solving the Sturmian secular equations (6.3), evaluating the matrix product on the right-hand side of (6.8), and making use of (6.4). (If the state Ψ_κ has a zero-field dipole moment, this must of course be taken into account.)

6.4 Induced transition dipole moments

If the two states Ψ_κ and $\Psi_{\kappa'}$ lie close together in energy, so that $p_\kappa \approx p_{\kappa'}$, then we can make the approximation

$$\int dx\ \Psi^*_\kappa(\mathbf{x}) \sum_{j=1}^{N} z_j \Psi_{\kappa'}(\mathbf{x}) \approx -\frac{1}{p_\kappa} \sum_{\nu,\nu'} B^*_{\nu,\kappa} T^{el.}_{\nu,\nu'} B_{\nu',\kappa'} \qquad (6.9)$$

In that case it is easy to calculate the field-induced transition dipole moment between the two states by the method just described. In the more general case, where this approximation is not valid, it is still possible (although more laborious) to calculate the induced transition dipole moment between two states. To do so, we begin by solving the secular equations (6.3) for many values of the parameter η. Then equation (6.4) can be used to make plots of $p_\kappa(\mathcal{E})$ and $p_{\kappa'}(\mathcal{E})$ for the two states in which we are interested. For a particular value of $\mathcal{E}$, we can find the corresponding values of the scaling parameters p_κ and $p_{\kappa'}$ for the two states in which we are interested. We next evaluate the matrix

$$V^{el.}_{\nu,\nu'} \equiv \int dx\ \Phi^*_\nu(\mathbf{x}, p_\kappa) \sum_{j=1}^{N} z_j\ \Phi_{\nu'}(\mathbf{x}, p_{\kappa'}) \qquad (6.10)$$

In equation (6.10) we include p_κ and $p_{\kappa'}$ as arguments of the two configurations, because they have different scaling parameters although they correspond to the same value of the applied field. Finally we calculate the induced transition dipole moment

$$\int dx\, \Psi_\kappa^*(\mathbf{x}) \sum_{j=1}^N z_j \Psi_{\kappa'}(\mathbf{x}) = \sum_{\nu,\nu'} B_{\nu,\kappa}^*(p_\kappa) V_{\nu,\nu'}^{el.} B_{\nu',\kappa'}(p_{\kappa'}) \tag{6.11}$$

where we have included the scaling factors as arguments in the matrix product on the right to remind ourselves that the B's do not correspond to the same value of η, although they correspond to the same value of $\mathcal{E}$.

6.5 External magnetic fields

We next turn to the spectral effect of magnetic fields. If one starts with the Dirac equation of an electron moving in an external electromagnetic potential, one can show that in the non-relativistic limit it leads to some extra terms that describe magnetic effects. When expressed in atomic units[2], where $e = \hbar = m = 1$ and $c = 137.036$, one of these terms has the form

$$v^{mag.}(\mathbf{x}_j) = \frac{1}{2c} \mathcal{H} \cdot (\mathbf{L}_j + 2\mathbf{S}_j) \tag{6.12}$$

For a system of N electrons, the corresponding term is

$$V^{mag.}(\mathbf{x}) = \frac{1}{2c} \sum_{j=1}^N \mathcal{H} \cdot (\mathbf{L}_j + 2\mathbf{S}_j) = \frac{1}{2c} \mathcal{H} \cdot (\mathbf{L} + 2\mathbf{S}) \tag{6.13}$$

Here $\mathcal{H}$ represents a uniform magnetic field vector, $\mathbf{L}$ is the total orbital angular momentum vector of the N-electron system, and $\mathbf{S}$ is the total spin vector. If the external magnetic field points in the direction of the z-axis of the system, this reduces to

$$V^{mag.}(\mathbf{x}) = \frac{\mathcal{H}_z}{2c}(L_z + 2S_z) \tag{6.14}$$

Remembering that generalized Sturmian configurations of the Goscinskian type are eigenfunctions of both L_z and S_z, we can write

$$T_{\nu',\nu}^{mag.} \equiv -\frac{1}{2c} \int dx\, \Phi_{\nu'}^*(\mathbf{x})(L_z + 2S_z)\Phi_\nu(\mathbf{x})$$

---

[2] 1 atomic unit of magnetic flux density is equal to $2.3505173 \cdot 10^5$ Tesla.

$$= -\frac{1}{2c}(M_L + 2M_S) \int dx\, \Phi_{\nu'}^*(\mathbf{x})\Phi_\nu(\mathbf{x}) \qquad (6.15)$$

If a constant external magnetic field in the z-direction is applied to the atom or ion, the secular equations become

$$\sum_\nu \left[\delta_{\nu',\nu} Z\mathcal{R} + T'_{\nu',\nu} + \zeta T_{\nu',\nu}^{mag.} - p_\kappa \delta_{\nu',\nu}\right] B_{\nu,\kappa} = 0 \qquad (6.16)$$

where

$$\zeta \equiv \frac{\mathcal{H}_z}{p_\kappa} \qquad (6.17)$$

The secular equations (6.16) can be solved for many values of the parameter ζ, thus generating a series of curves $p_\kappa(\zeta)$. Then with the help of equations (6.17) and the relationship $E_\kappa = -p_\kappa^2/2$, we can construct curves representing $E_\kappa(\mathcal{H}_z)$, as is illustrated in Figure 6.3.

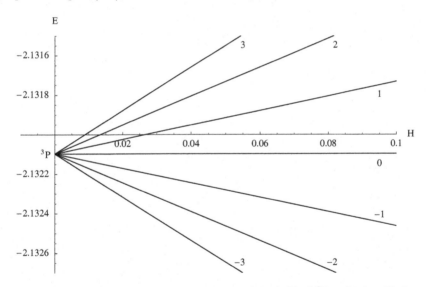

Fig. 6.3 This figure shows the magnetic splitting of the (1s)(2p) ^{3}P multiplet of helium. Here spin-orbit coupling is neglected, but it will be discussed in Chapter 7. Each energy level corresponds to a value of $M_L + 2M_S$ where $M_L \in \{-1, 0, 1\}$ and $2M_S \in \{-2, 0, 2\}$. The levels corresponding to $M_L + 2M_S = \pm 1$ are 2-fold degenerate, while the remaining levels are non-degenerate.

Chapter 7

RELATIVISTIC EFFECTS

7.1 Lorentz invariance and 4-vectors

Albert Einstein's special theory of relativity was built on the negative result of the Michaelson-Morley experiment, an experiment that attempted to measure the absolute velocity of the earth through space. Einstein boldly postulated that no experiment whatever can measure absolute motion, that is to say, according to his postulate it is impossible for an observer to know whether he is in a state of rest or in a state of uniform motion. All inertial frames are equivalent. Einstein's postulate has been amply confirmed by experiment, and today it is one of the basic principles of modern physics.

The equivalence of all inertial frames can be expressed in another way: Every fundamental physical law must exhibit symmetry between the space and time coordinates in such a way that ict enters on the same footing as the Cartesian coordinates x, y and z. (Here $i \equiv \sqrt{-1}$, while c is the velocity of light, and t is the time.) In relativistic theory, space and time combine to form a pseudo-Euclidean space-time continuum (Minkowski space). A transformation from one inertial frame to another (a Lorentz transformation) corresponds to a rotation in this space, and such a transformation must leave all fundamental physical laws invariant in form.

Every physical quantity that is represented by a 3-component vector in non-relativistic theory has a 4th component in the relativistic 4-dimensional space-time continuum. Thus, for example, the position vector $\mathbf{x} = (x, y, z)$ in 3-dimensional space has a 4th component in relativistic theory:

$$x_\lambda = (x, y, z, ict) = (\mathbf{x}, ict) \tag{7.1}$$

while the vector potential $\mathbf{A} = (A_x, A_y, A_z)$ in electromagnetic theory is the space component of a 4-vector, whose 4th component is i multiplied by the electrostatic potential ϕ:

$$A_\lambda = (A_x, A_y, A_z, i\phi) = (\mathbf{A}, i\phi) \tag{7.2}$$

Similarly, the current density vector $\mathbf{j} = (j_x, j_y, j_z)$ is the space-component of a 4-vector

$$j_\lambda = (j_x, j_y, j_z, ic\rho) = (\mathbf{j}, ic\rho) \qquad (7.3)$$

whose time-component is ic multiplied by the charge density ρ. (Throughout this chapter we will represent 3-vectors by writing them in bold-face letters. Thus $j_\lambda = (\mathbf{j}, ic\rho)$ means that the first three components of the 4-vector j_λ are given by $\mathbf{j} = (j_x, j_y, j_z)$, while the 4th component is $ic\rho$.) The gradient operator $\boldsymbol{\partial}$ also becomes the space-component of a 4-vector in relativistic theory:

$$\partial_\lambda \equiv \left(\frac{\partial}{\partial x_1}, \frac{\partial}{\partial x_2}, \frac{\partial}{\partial x_3}, \frac{\partial}{\partial x_4}\right) = \left(\boldsymbol{\partial}, -\frac{i}{c}\frac{\partial}{\partial t}\right) \qquad (7.4)$$

while the Laplacian operator is replaced by the d'Alembertian operator:

$$\Box \equiv \sum_{\lambda=1}^{4} \partial_\lambda^2 \equiv \sum_{\lambda=1}^{4} \frac{\partial^2}{\partial x_\lambda} \qquad (7.5)$$

an operator which exhibits the required space-time symmetry, so that its form is the same in all inertial frames. In relativistic electrodynamics, the electric field vector $\boldsymbol{\mathcal{E}}$ and the magnetic field vector $\boldsymbol{\mathcal{H}}$ are components of an antisymmetric tensor $F_{\lambda',\lambda}$, which is related to A_λ by

$$F_{\lambda',\lambda} \equiv \partial_{\lambda'} A_\lambda - \partial_\lambda A_{\lambda'} = \begin{pmatrix} 0 & \mathcal{H}_z & -\mathcal{H}_y & -i\mathcal{E}_x \\ -\mathcal{H}_z & 0 & \mathcal{H}_x & -i\mathcal{E}_y \\ \mathcal{H}_y & -\mathcal{H}_x & 0 & -i\mathcal{E}_z \\ i\mathcal{E}_x & i\mathcal{E}_y & i\mathcal{E}_z & 0 \end{pmatrix} \qquad (7.6)$$

The 4-vector A_λ, which represents the electromagnetic potential, is related to the 4-vector representing current density by

$$\Box A_\lambda = -\frac{4\pi}{c} j_\lambda \qquad (7.7)$$

When both the current density j_λ and the electromagnetic potential 4-vector A_λ are independent of time, equation (7.7) reduces to:

$$\nabla_1^2 A_\lambda(\mathbf{x}_1) = -\frac{4\pi}{c} j_\lambda(\mathbf{x}_1) \qquad (7.8)$$

which has the Green's function solution

$$A_\lambda(\mathbf{x}_1) = \frac{1}{c} \int d^3 x_2 \, \frac{1}{|\mathbf{x}_1 - \mathbf{x}_2|} \, j_\lambda(\mathbf{x}_2) \qquad (7.9)$$

We can see that (7.9) is a solution to (7.8) because

$$\nabla_1^2 \frac{1}{|\mathbf{x}_1 - \mathbf{x}_2|} = -4\pi \delta^3(\mathbf{x}_1 - \mathbf{x}_2) \equiv -4\pi \delta(x_1 - x_2)\delta(y_1 - y_2)\delta(z_1 - z_2) \quad (7.10)$$

and therefore

$$\begin{aligned}
\nabla_1^2 A_\lambda(\mathbf{x}_1) &= \frac{1}{c} \int d^3 x_2 \, \nabla_1^2 \frac{1}{|\mathbf{x}_1 - \mathbf{x}_2|} \, j_\lambda(\mathbf{x}_2) \\
&= -\frac{4\pi}{c} \int d^3 x_2 \, \delta^3(\mathbf{x}_1 - \mathbf{x}_2) \, j_\lambda(\mathbf{x}_2) \\
&= -\frac{4\pi}{c} j_\lambda(\mathbf{x}_1)
\end{aligned} \quad (7.11)$$

The subscript 1 on the Laplacian operator means that the operator is acting on the coordinates of the field-point $\mathbf{x}_1$ rather than on the source-point, $\mathbf{x}_2$.

Because of charge conservation, the current density 4-vector obeys the condition

$$\sum_{\lambda=1}^{4} \partial_\lambda j_\lambda = 0 \quad (7.12)$$

Since the current density is related to the electromagnetic potential 4-vector through (7.7), it is natural to work in the Lorentz gauge, where a similar condition is imposed on A_λ:

$$\sum_{\lambda=1}^{4} \partial_\lambda A_\lambda = 0 \quad (7.13)$$

Equations (7.7) and (7.13) are Maxwell's equations in a vacuum, written in a form that makes the space-time symmetry apparent.

7.2 The Dirac equation for an electron in an external electromagnetic potential

P.A.M. Dirac's relativistic wave equation for an electron moving in an external potential A_λ can be written in the form:

$$\left[\sum_{\lambda=1}^{4} \gamma_\lambda \left(\partial_\lambda - \frac{i}{c} A_\lambda \right) + c \right] \chi_\mu = 0 \quad (7.14)$$

where atomic units are used and where the γ_λ's are 4×4 matrices:

$$\gamma_1 = \begin{pmatrix} 0 & 0 & 0 & -i \\ 0 & 0 & -i & 0 \\ 0 & i & 0 & 0 \\ i & 0 & 0 & 0 \end{pmatrix} \qquad \gamma_2 = \begin{pmatrix} 0 & 0 & 0 & -1 \\ 0 & 0 & 1 & 0 \\ 0 & 1 & 0 & 0 \\ -1 & 0 & 0 & 0 \end{pmatrix} \qquad (7.15)$$

$$\gamma_3 = \begin{pmatrix} 0 & 0 & -i & 0 \\ 0 & 0 & 0 & i \\ i & 0 & 0 & 0 \\ 0 & -i & 0 & 0 \end{pmatrix} \qquad \gamma_4 = \begin{pmatrix} 1 & 0 & 0 & 0 \\ 0 & 1 & 0 & 0 \\ 0 & 0 & -1 & 0 \\ 0 & 0 & 0 & -1 \end{pmatrix} \qquad (7.16)$$

In atomic units, the electron rest-mass is equal to 1, and Planck's constant divided by 2π is also equal to 1, while the velocity of light has a value equal to the reciprocal of the fine structure constant:

$$m_0 = 1 \qquad \hbar = 1 \qquad c = 137.036 \qquad (7.17)$$

From the definitions of the γ_λ's, it follows that they anticommute:

$$\gamma_{\lambda'}\gamma_\lambda + \gamma_\lambda\gamma_{\lambda'} = 2I\delta_{\lambda',\lambda} \qquad (7.18)$$

In equation (7.18), I is a 4×4 unit matrix. Solutions to the 1-electron Dirac equation are 4-component spinors.

7.3 Time-independent problems

In the special case where the external electromagnetic potential 4-vector A_λ is independent of time, it is convenient to write the Dirac equation (7.14) in a different form, where we introduce the notation

$$\boldsymbol{\alpha} = i\gamma_0\boldsymbol{\gamma} \qquad \gamma_0 \equiv \gamma_4 \qquad (7.19)$$

From equations (7.15), (7.16) and (7.19) it follows that the components of the 3-vector $\boldsymbol{\alpha}$ can be written in block form as

$$\alpha_j = \left(\begin{array}{c|c} 0 & \sigma_j \\ \hline \sigma_j & 0 \end{array}\right) \qquad j = 1, 2, 3 \qquad (7.20)$$

where, in the off-diagonal blocks, σ_j, $j = 1, 2, 3$ are the 2×2 Pauli spin matrices:

$$\sigma_1 = \begin{pmatrix} 0 & 1 \\ 1 & 0 \end{pmatrix}, \qquad \sigma_2 = \begin{pmatrix} 0 & -i \\ i & 0 \end{pmatrix}, \qquad \sigma_3 = \begin{pmatrix} 1 & 0 \\ 0 & -1 \end{pmatrix} \qquad (7.21)$$

For time-independent problems, the Dirac equation for a single electron can then be written in the form:
$$[H - \epsilon_\mu]\chi_\mu(\mathbf{x}) = 0 \tag{7.22}$$

where
$$H = -ic\boldsymbol{\alpha} \cdot \left(\boldsymbol{\partial} - \frac{i}{c}\mathbf{A}(\mathbf{x})\right) + I\phi(\mathbf{x}) + \gamma_0 c^2 \tag{7.23}$$

is the Dirac Hamiltonian of an electron moving in a constant external electromagnetic potential, ϵ_μ is the 1-electron energy, and $\chi_\mu(\mathbf{x})$ is the 4-component time-independent spinor of the electron. The kinetic energy term in the Dirac Hamilton is given by

$$-ic\boldsymbol{\alpha} \cdot \boldsymbol{\partial} = -ic\begin{pmatrix} 0 & 0 & \partial_3 & \partial_- \\ 0 & 0 & \partial_+ & -\partial_3 \\ \partial_3 & \partial_- & 0 & 0 \\ \partial_+ & -\partial_3 & 0 & 0 \end{pmatrix} \tag{7.24}$$

where
$$\partial_\pm \equiv \partial_1 \pm i\partial_2 \tag{7.25}$$

Similarly, the part of the Dirac Hamiltonian involving potentials is

$$-\boldsymbol{\alpha} \cdot \mathbf{A} + I\phi = \begin{pmatrix} \phi & 0 & -A_3 & -A_- \\ 0 & \phi & -A_+ & A_3 \\ -A_3 & -A_- & \phi & 0 \\ -A_+ & A_3 & 0 & \phi \end{pmatrix} \tag{7.26}$$

where
$$A_\pm \equiv A_1 \pm iA_2 \tag{7.27}$$

7.4 The Dirac equation for an electron in the field of a nucleus

When $\mathbf{A}(\mathbf{x}) = 0$, and $\phi(\mathbf{x}) = -Z/r$, equation (7.22) reduces to

$$\left[-ic\boldsymbol{\alpha} \cdot \boldsymbol{\partial} - \frac{Z}{r} + \gamma_0 c^2 - \epsilon_\mu\right]\chi_\mu(\mathbf{x}) = 0 \tag{7.28}$$

which is the Dirac equation for an electron moving in the attractive electrostatic potential of a nucleus with charge Z. Equation (7.28) can be solved exactly, and the solutions have the form

$$\chi_\mu(\mathbf{x}) = \chi_{njlM}(\mathbf{x}) = \begin{pmatrix} ig_{njl}(r)\Omega_{j,l,M}(\theta,\varphi) \\ -f_{njl}(r)\Omega_{j,2j-l,M}(\theta,\varphi) \end{pmatrix} \tag{7.29}$$

Examples are shown in equations (7.54) and (7.54). In equation (7.29), the angular function $\Omega_{j,l,M}(\theta,\varphi)$ is a two-component "spherical spinor", which is an eigenfunction of orbital angular momentum corresponding to the quantum number l, total angular momentum (orbital plus spin) with quantum number j, and the z-component of total angular momentum, with quantum number M. The spherical spinors are built up from spherical harmonics and 2-component spinors by combining them with the appropriate Clebsch-Gordan coefficients in such a way as to produce eigenfunctions of total angular momentum. The Clebsch-Gordan coefficients that enter are different, depending on whether $j = l + \frac{1}{2}$ or $j = l - \frac{1}{2}$. When $j = l + \frac{1}{2}$,

$$\Omega_{j,l,M}(\theta,\varphi) = \begin{pmatrix} \sqrt{\dfrac{l+M+\frac{1}{2}}{2l+1}}\, Y_{l,M-\frac{1}{2}}(\theta,\varphi) \\ \sqrt{\dfrac{l-M+\frac{1}{2}}{2l+1}}\, Y_{l,M+\frac{1}{2}}(\theta,\varphi) \end{pmatrix} \tag{7.30}$$

while when $j = l - \frac{1}{2}$,

$$\Omega_{j,l,M}(\theta,\varphi) = \begin{pmatrix} -\sqrt{\dfrac{l-M+\frac{1}{2}}{2l+1}}\, Y_{l,M-\frac{1}{2}}(\theta,\varphi) \\ \sqrt{\dfrac{l+M+\frac{1}{2}}{2l+1}}\, Y_{l,M+\frac{1}{2}}(\theta,\varphi) \end{pmatrix} \tag{7.31}$$

The radial function $g_{njl}(r)$ is much larger than $f_{njl}(r)$. The large and small radial functions are defined respectively by

$$g_{njl}(r) = \mathcal{N} r^{\gamma-1} e^{-Zr/\bar{n}} \left(W_1(r) - W_2(r)\right) \tag{7.32}$$

and

$$f_{njl}(r) = \mathcal{N} \sqrt{\frac{c^2 - \epsilon_{nj}}{c^2 + \epsilon_{nj}}}\, r^{\gamma-1} e^{-Zr/\bar{n}} \left(W_1(r) + W_2(r)\right) \tag{7.33}$$

where

$$W_1(r) \equiv n_r F\left(j + \frac{1}{2} - n + 1 \Big| 2\gamma + 1 \Big| \frac{2Zr}{\bar{n}}\right)$$

$$W_2(r) \equiv (\bar{n} - \kappa) F\left(j + \frac{1}{2} - n \Big| 2\gamma + 1 \Big| \frac{2Zr}{\bar{n}}\right) \tag{7.34}$$

with

$$\kappa \equiv \begin{cases} -(j + \frac{1}{2}) & j = l + \frac{1}{2} \\ j + \frac{1}{2} & j = l - \frac{1}{2} \end{cases} \tag{7.35}$$

$$\gamma \equiv \sqrt{\left(j + \frac{1}{2}\right)^2 - \left(\frac{Z}{c}\right)^2} \tag{7.36}$$

$$n_r \equiv n - j - \frac{1}{2} \tag{7.37}$$

and

$$\bar{n} \equiv \sqrt{n^2 - 2n_r(j + \frac{1}{2} - \gamma)} \tag{7.38}$$

Just as in the definition of the non-relativistic hydrogenlike orbitals, $F(a|b|\zeta)$ is a confluent hypergeometric function:

$$F(a|b|\zeta) \equiv 1 + \frac{a}{b}\zeta + \frac{a(a+1)}{b(b+1)2!}\zeta^2 + \cdots \tag{7.39}$$

When $Z \ll 137$, the 1-electron energies

$$\epsilon_{nj} = \frac{c^2}{\sqrt{1 + \left(\frac{Z}{c(\gamma+n-|j+\frac{1}{2}|)}\right)^2}} \tag{7.40}$$

are only slightly smaller than c^2, and hence $c^2 - \epsilon_{nj}$ is much smaller than

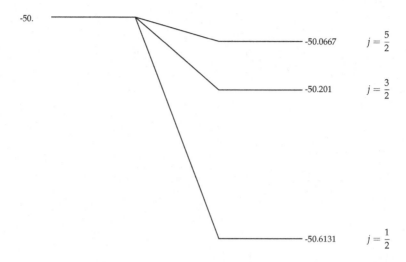

Fig. 7.1 This figure shows the relativistic splitting of the $n=3$ level of a hydrogenlike atom with nuclear charge $Z=30$. The relativistic energies depend on j, and the binding energy is larger than in the non-relativistic case.

$c^2 + \epsilon_{nj}$. For this reason

$$\sqrt{\frac{c^2 - \epsilon_{nj}}{c^2 + \epsilon_{nj}}} \ll 1 \tag{7.41}$$

and the radial functions $f_{njl}(r)$, which contain this factor, are considerably smaller than $g_{njl}(r)$. The normalizing constant that appears in the definition of the radial functions is given by

$$\mathcal{N} \equiv \frac{1}{\Gamma(2\gamma+1)} \sqrt{\frac{\Gamma(2\gamma+n_r+1)(1+\epsilon_{nj}/c^2)}{4n_r! \bar{n}(\bar{n}-\kappa)}} \left(\frac{2Z}{\bar{n}}\right)^{\gamma+\frac{1}{2}} \tag{7.42}$$

where

$$\Gamma(z) \equiv \int_0^\infty dt\, e^{-t} t^{z-1} \tag{7.43}$$

The hydrogenlike Dirac spinors $\chi_\mu(\mathbf{x})$ obey the orthonormality relation

$$\int d\tau\, \chi_{\mu'}^\dagger(\mathbf{x}) \chi_\mu(\mathbf{x}) = \delta_{\mu',\mu} \tag{7.44}$$

where the dagger means "conjugate transpose", and where $\int d\tau$ implies both a summation over spinor indices and an integration over space coordinates.

RELATIVISTIC EFFECTS

Table 7.1: This table compares the non-relativistic and relativistic calculated ground state energies of the 1-electron isoelectronic series with experimental values from the NIST tables. The calculated values include corrections for the motion of the nucleus. As Z increases, relativistic effects become more important and motion of the nucleus less so. At large Z, the non-relativistic calculated energies greatly underestimate the binding energies. The small differences between experimental values and relativistic energies calculated from equation (7.45) are due to quantum electrodynamic effects, such as the Lamb shift and vacuum polarization.

		$n=1$	$n=2$	$n=3$
	Non-rel.	-0.499728	-0.124932	-0.055525
$Z=1$	Rel.	-0.499734	-0.124934	-0.055556
	Exp.	-0.499733	-0.124933	-0.055526
	Non-rel.	-49.99863	-12.49966	-5.555404
$Z=10$	Rel.	-50.06538	-12.52052	-5.562819
	Exp.	-50.05988	-12.51979	-5.562603
	Non-rel.	-199.9973	-49.99932	-22.22192
$Z=20$	Rel.	-201.0738	-50.33591	-22.34147
	Exp.	-201.0137	-50.32775	-22.33893
	Non-rel.	-449.9959	-112.4990	-49.99954
$Z=30$	Rel.	-455.5208	-114.2276	-50.61267
	Exp.	-455.2848	-114.1946	-50.60222

One can ask how well the calculated relativistic energies given by equation (7.40) agree with the experimental energies for the 1-electron isoelectronic series. Remembering that in atomic units the rest-mass of the electron is equal to 1, we have

$$\epsilon_{nj}(Z) - c^2 = c^2 \left(\frac{1}{\sqrt{1 + \left(\frac{Z}{c(\gamma+n-|j+\frac{1}{2}|)}\right)^2}} - 1 \right) \qquad c = 137.036 \quad (7.45)$$

which gives us the calculated relativistic 1-electron energies in Hartrees as a function of the nuclear charge Z. Values of the experimental and calculated energies for various values of Z and n (with $j = 1/2$) are shown in Table 7.1 and Figure 7.2.

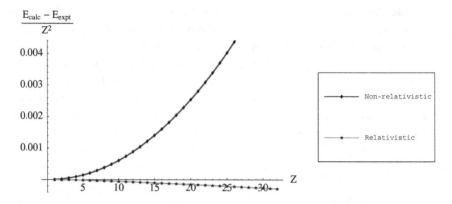

Fig. 7.2 This figure shows the difference between the experimental and the calculated ground state energies for the 1-electron isoelectronic series in Hartrees, divided by Z^2. The calculated energies include corrections for the motion of the nucleus. As Z increases, the non-relativistic calculation show the electron to be much less tightly bound than the spectroscopic experiments. The relativistic calculation at high Z slightly overestimates the binding energy because of quantum electrodynamic effects, such as vacuum polarization and the Lamb shift.

7.5 Relativistic formulation of the Zeeman and Paschen-Bach effects

If we introduce the definition

$$\tilde{\Omega}_\mu(\mathbf{u}) \equiv \tilde{\Omega}_{j,l,M}(\mathbf{u}) \equiv \Omega_{j,2j-l,M}(\mathbf{u}) \tag{7.46}$$

then equation (7.29) can be rewritten in the condensed form

$$\chi_\mu(\mathbf{x}) = \chi_{njlM}(\mathbf{x}) \equiv \begin{pmatrix} ig_\mu(r)\Omega_\mu(\mathbf{u}) \\ -f_\mu(r)\tilde{\Omega}_\mu(\mathbf{u}) \end{pmatrix} \tag{7.47}$$

while the conjugate spin-orbitals become

$$\chi^\dagger_{\mu'}(\mathbf{x}) = \left(-ig_{\mu'}(r)\Omega^\dagger_{\mu'}(\mathbf{u}),\ -f_{\mu'}(r)\tilde{\Omega}^\dagger_{\mu'}(\mathbf{u})\right) \tag{7.48}$$

If we define the transition current vector to be the quantity

$$\mathbf{j}_{\mu',\mu} \equiv c(\chi^\dagger_{\mu'}\boldsymbol{\alpha}\chi_\mu) \tag{7.49}$$

then

$$\mathbf{j}_{\mu',\mu} = icg_{\mu'}f_\mu(\Omega^\dagger_{\mu'}\boldsymbol{\sigma}\tilde{\Omega}_\mu) - icf_{\mu'}g_\mu(\tilde{\Omega}^\dagger_{\mu'}\boldsymbol{\sigma}\Omega_\mu) \tag{7.50}$$

RELATIVISTIC EFFECTS

where we have made use of equations (7.20), (7.21) and (7.46)-(7.49). From equation (7.26) it follows that the part of the Hamiltonian due to an external vector potential $\mathbf{A}$ is

$$v^{mag} = -\boldsymbol{\alpha} \cdot \mathbf{A} \tag{7.51}$$

A constant magnetic field in the z-direction

$$\mathcal{H} = \boldsymbol{\partial} \times \mathbf{A} = (0, 0, \mathcal{H}) \tag{7.52}$$

can be represented by the vector potential

$$\mathbf{A} = \frac{\mathcal{H}}{2}(-y, x, 0) \tag{7.53}$$

The matrix element of v^{mag} based on the relativistic hydrogenlike spin-orbitals will then be given by

$$v^{mag}_{\mu',\mu} \equiv -\int d^3x \, (\chi^\dagger_{\mu'} \boldsymbol{\alpha} \cdot \mathbf{A} \chi_\mu) = -\frac{1}{c} \int d^3x \, \mathbf{j}_{\mu',\mu} \cdot \mathbf{A} \tag{7.54}$$

which can be rewritten in the form

$$v^{mag}_{\mu',\mu} = \int d^3x \left[-ig_{\mu'} f_\mu (\Omega^\dagger_{\mu'} \boldsymbol{\sigma} \cdot \mathbf{A} \tilde{\Omega}_\mu) + if_{\mu'} g_\mu (\tilde{\Omega}^\dagger_{\mu'} \boldsymbol{\sigma} \cdot \mathbf{A} \Omega_\mu) \right] \tag{7.55}$$

where, from equations (7.20), (7.21) and (7.53), it follows that

$$\boldsymbol{\sigma} \cdot \mathbf{A} = \frac{i\mathcal{H}}{2} \begin{pmatrix} 0 & -x_- \\ x_+ & 0 \end{pmatrix} \tag{7.56}$$

with $x_+ \equiv x+iy$ and $x_- \equiv x-iy$. We can calculate the effect of an external magnetic field on an electron moving in the field of a nucleus of charge Z by constructing a matrix representation of the total Dirac Hamiltonian, including both the nuclear attraction potential and the external magnetic field. If we base this representation on the relativistic hydrogenlike spin-orbitals,

$$H_0 \equiv -ic\boldsymbol{\alpha} \cdot \boldsymbol{\partial} - \frac{Z}{r} + \gamma_0 c^2 \tag{7.57}$$

will already be diagonal, with diagonal elements given by equation (7.40). To this diagonal matrix, $v^{mag}_{\mu',\mu}$ must be added, and the total matrix rediagonalized. In other words, the energies and wave functions for an electron moving in the combined potential of a nucleus and an external magnetic field can be found by diagonalizing the matrix

$$H_{\mu',\mu} = \delta_{\mu',\mu} \epsilon_\mu + v^{mag}_{\mu',\mu} \tag{7.58}$$

where ϵ_μ is given by equation (7.40).

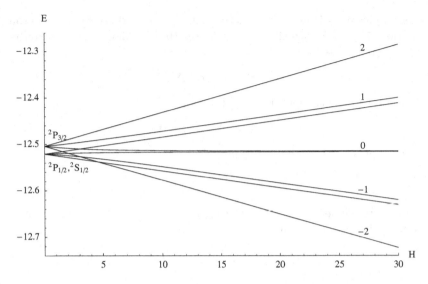

Fig. 7.3 This figure illustrates the effect of a magnetic field of increasing strength acting on the $n = 2$ states of N^{9+}. In the absence of a magnetic field, solution of the Dirac equation in the field of the nucleus yields two sets of fourfold degenerate states, corresponding respectively to $j = \frac{1}{2}$ and $j = \frac{3}{2}$. The calculated splitting between these two sets of states is $\epsilon_{2,\frac{3}{2}} - \epsilon_{2,\frac{1}{2}} = 0.016697$ Hartree, which is close to the observed value of 0.016733 Hartree. On the left hand side of the figure, the fourfold degenerate states are linearly split by the magnetic field. This is the normal Zeeman effect. When the effect of the magnetic field becomes large compared to the spin-orbit coupling, the energies depend on $m + 2m_S$. The strong field case is sometimes called the Paschen-Bach effect.

7.6 Relativistic many-electron Sturmians

Let us now attempt to adapt the generalized Sturmian method to a system of N electrons for which relativistic effects are important. We can begin by introducing the definitions

$$D_0 \equiv -ic \sum_{k=1}^{N} \boldsymbol{\alpha}_k \cdot \boldsymbol{\partial}_k \qquad (7.59)$$

and

$$V_0 \equiv -\sum_{k=1}^{N} \frac{Z}{r_k} \qquad (7.60)$$

RELATIVISTIC EFFECTS

In equations (7.59) and (7.60), the subscript k indicates that the operators act on the Dirac spinor of the kth electron. We also let

$$\beta_\nu \equiv \frac{Q_\nu}{Z} \qquad (7.61)$$

Then, if we construct a Slater determinant from the hydrogenlike Dirac spinors discussed above with Z replaced by Q_ν, the Slater determinant

$$\Phi_\nu = \mathcal{N}_\nu |\chi_\mu \chi_{\mu'} \chi_{\mu''} \cdots| \qquad (7.62)$$

will be an exact solution to the many-electron equation

$$\left[D_0 + \beta_\nu V_0 - E_\kappa - Nc^2 \right] \Phi_\nu = 0 \qquad (7.63)$$

provided that

$$\epsilon_\mu + \epsilon_{\mu'} + \epsilon_{\mu''} + \cdots + = E_\kappa + Nc^2 \qquad (7.64)$$

In equation (7.62), $\mathcal{N}_\nu$ is a normalizing constant. Equation (7.63) is analogous to the approximate many-electron Schrödinger equation (3.4), and the Dirac configurations of equation (7.62) are the relativistic analogue of the Goscinskian configurations. E_κ is the binding energy of the N-electron atom, while Nc^2 represents the total rest energy of the electrons in Hartrees.

Just as in the non-relativistic case, it is possible to establish a potential-weighted orthonormality relation between the configurations. If both Φ_ν and $\Phi_{\nu'}$ are solutions to (7.63), then

$$\int d\tau \, \Phi_{\nu'}^\dagger \left[D_0 - E_\kappa - Nc^2 \right] \Phi_\nu = -\beta_\nu \int d\tau \, \Phi_{\nu'}^\dagger V_0 \Phi_\nu \qquad (7.65)$$

and

$$\int d\tau \, \Phi_\nu^\dagger \left[D_0 - E_\kappa - Nc^2 \right] \Phi_{\nu'} = -\beta_{\nu'} \int d\tau \, \Phi_\nu^\dagger V_0 \Phi_{\nu'} \qquad (7.66)$$

Subtracting the conjugate transpose of (7.66) from (7.65), and making use of Hermiticity, we have

$$(\beta_{\nu'} - \beta_\nu) \int d\tau \, \Phi_{\nu'}^\dagger V_0 \Phi_\nu = 0 \qquad (7.67)$$

so that

$$\int d\tau \, \Phi_{\nu'}^\dagger V_0 \Phi_\nu = 0 \quad \text{if } \beta_{\nu'} \neq \beta_\nu \qquad (7.68)$$

It is convenient to normalize the Dirac configurations in such a way that they obey the potential-weighted orthonormality relations

$$\int d\tau\, \Phi_{\nu'}^\dagger V_0 \Phi_\nu = \frac{2E_\kappa}{\beta_\nu} \delta_{\nu',\nu} \qquad (7.69)$$

Just as in the non-relativistic case, orthogonality of ϕ_ν and $\phi_{\nu'}$ when $\beta_\nu = \beta_{\nu'}$ but $\nu \neq \nu'$ follows from the ordinary Slater-Condon rules because orthonormality of the radial functions then holds, and because ν and ν' must differ by at least two μ's.

We are now in a position to construct solutions to the relativistic many-particle wave equation

$$\left[D_0 + V_0 + V' - E_\kappa - Nc^2 \right] \Psi_\kappa = 0 \qquad (7.70)$$

where V' represents the Gaunt interaction operator[1]

$$V' = \sum_{k',k=1}^{N} \sum_{\lambda=1}^{4} \frac{(\gamma_0 \gamma_\lambda)_{k'} (\gamma_0 \gamma_\lambda)_k}{|\mathbf{x}_k - \mathbf{x}_{k'}|} = \sum_{k',k=1}^{N} \frac{I_{k'} I_k - \boldsymbol{\alpha}_{k'} \cdot \boldsymbol{\alpha}_k}{|\mathbf{x}_k - \mathbf{x}_{k'}|} \qquad (7.71)$$

In equation (7.71), I_k is a 4×4 identity matrix that acts on the spinor of electron k. If we represent the N-electron wave function Ψ_κ as a superposition of Dirac configurations,

$$\Psi_\kappa = \sum_\nu \Phi_\nu B_{\kappa,\nu} \qquad (7.72)$$

we obtain

$$\sum_\nu \left[D_0 + V_0 + V' - E_\kappa - Nc^2 \right] \Phi_\nu B_{\kappa,\nu} = 0 \qquad (7.73)$$

Multiplying (7.73) from the left by a conjugate configuration from the basis set and integrating over all space and spin coordinates yields

$$\sum_\nu \int d\tau\, \Phi_{\nu'}^\dagger \left[D_0 + V_0 + V' - E_\kappa - Nc^2 \right] \Phi_\nu B_{\kappa,\nu} = 0 \qquad (7.74)$$

Since all of the Dirac configurations in the basis set are solutions to equation (7.63), (7.74) can be rewritten in the form

$$\sum_\nu \int d\tau\, \Phi_{\nu'}^\dagger \left[V_0 + V' - \beta_\nu V_0 \right] \Phi_\nu B_{\kappa,\nu} = 0 \qquad (7.75)$$

---

[1] The Gaunt interaction [Gaunt, 1929] neglects the retardation effects that ought to be included for exchange terms (see, for example, [Bethe and Salpeter, 1977]). However, neglecting retardation is a good approximation for interelectron interactions within atoms.

Making use of the generalized Slater-Condon rules, we can write the matrix element of the interelectron interaction operator in the form

$$V'_{\nu',\nu} \equiv \int d\tau \, \Phi^\dagger_{\nu'} V' \Phi_\nu = \mathcal{N}_{\nu'} \mathcal{N}_\nu \sum_{i=1}^{N} \sum_{j=i+1}^{N} \sum_{k=1}^{N} \sum_{l=k+1}^{N} (-1)^{i+j+k+l} C_{ij;kl} |S_{ij;kl}| \tag{7.76}$$

where

$$C_{ij;kl} \equiv -\frac{1}{c^2} \int d^3x_1 \int d^3x_2 \sum_{\lambda=1}^{4} \frac{\left[j^{ij}_\lambda(\mathbf{x}_1) j^{kl}_\lambda(\mathbf{x}_2) - j^{ik}_\lambda(\mathbf{x}_1) j^{jl}_\lambda(\mathbf{x}_2) \right]}{|\mathbf{x}_1 - \mathbf{x}_2|} \tag{7.77}$$

and where $|S_{ij;kl}|$ is the determinant of the matrix formed from the matrix S of overlap integrals by deleting the i'th and j'th row and the k'th and l'th columns. The 4-current transition densities appearing in (7.77) are given by

$$j^{\mu',\mu}_\lambda(\mathbf{x}_1) = ic(\chi^\dagger_{\mu'}(\mathbf{x}_1)(\gamma_0 \gamma_\lambda)_1 \chi_\mu(\mathbf{x}_1)) \tag{7.78}$$

which can be rewritten in the form

$$j^{\mu',\mu}_\lambda(\mathbf{x}_1) = (\mathbf{j}_{\mu',\mu}(\mathbf{x}_1), ic\rho_{\mu',\mu}(\mathbf{x}_1)) \tag{7.79}$$

where

$$\mathbf{j}_{\mu',\mu}(\mathbf{x_k}) = c(\chi^\dagger_{\mu'}(\mathbf{x_k}) \boldsymbol{\alpha}_k \chi_\mu(\mathbf{x_k})) \tag{7.80}$$

and

$$ic\rho_{\mu',\mu}(\mathbf{x_k}) = ic(\chi^\dagger_{\mu'}(\mathbf{x_k}) \chi_\mu(\mathbf{x_k})) \tag{7.81}$$

Making use of the potential-weighted orthonormality relation (7.69), we can rewrite (7.74) in the form:

$$\sum_\nu \left[\frac{2E_\kappa}{\beta_\nu} \delta_{\nu',\nu} + V'_{\nu',\nu} - 2E_\kappa \delta_{\nu',\nu} \right] B_{\kappa,\nu} = 0 \tag{7.82}$$

Just as in the non-relativistic case, we can introduce the definition:

$$p_\kappa \equiv \sqrt{-2E_\kappa} \tag{7.83}$$

in terms of which (7.82) becomes

$$\sum_\nu \left[-\frac{p_\kappa^2}{\beta_\nu} \delta_{\nu',\nu} + V'_{\nu',\nu} + p_\kappa^2 \delta_{\nu',\nu} \right] B_{\kappa,\nu} = 0 \tag{7.84}$$

Finally, changing all signs, dividing by p_κ, and making use of (7.61), we obtain

$$\sum_\nu \left[Z \frac{p_\kappa}{Q_\nu} \delta_{\nu',\nu} + T'_{\nu',\nu} - p_\kappa \delta_{\nu',\nu} \right] B_{\kappa,\nu} = 0 \qquad (7.85)$$

where

$$T'_{\nu',\nu} \equiv -\frac{1}{p_\kappa} V'_{\nu',\nu} \qquad (7.86)$$

Equation (7.85) has the form of the non-relativistic set of secular equations (3.39), except that $\mathcal{R}_\nu$ has been replaced by p_κ/Q_ν. Some of the neatness of the non-relativistic case is lost, because both p_κ/Q_ν and $T'_{\nu',\nu}$ are energy-dependent (although only weakly so). Nevertheless, the relativistic generalized Sturmian problem can be solved by the following procedure:

(1) **Compute effective charges Q_ν as functions of p_κ:**
Tabulate (p_κ, Q_ν)-pairs for a range of p_κ. In order that the Dirac configurations will satisfy equation (7.63), the Q_ν must be chosen such that

$$p_\kappa(Q_\nu) = \sqrt{2\left(Nc^2 - \sum_{\mu \in \nu} \epsilon_\mu(Q_\nu)\right)} \qquad (7.87)$$

Since the above expression is monotonic in Q_ν, interpolating the points will give Q_ν as a function of p_κ. The curvature is very small, so the approximation is good even with few points.

(2) **Compute normalization factors $\mathcal{N}_\nu$ as functions of p_κ:**
To satisfy (7.69), the normalization factor $\mathcal{N}_\nu$ for Φ_ν must be chosen to be

$$\mathcal{N}_\nu = \frac{p_\kappa}{\sqrt{Q_\nu I_\nu}} \qquad (7.88)$$

where

$$I_\nu = \int_0^\infty dr \, r^2 \sum_{\mu \in \nu} \left(\frac{f_\mu(r)^2 + g_\mu(r)^2}{r} \right) \qquad (7.89)$$

One may interpolate a table of $(p_\kappa, \mathcal{N}_\nu)$-pairs to obtain $\mathcal{N}_\nu$ as a function of p_κ, in much the same way that interpolation was used in step (1).

RELATIVISTIC EFFECTS

(3) **Obtain approximate $\bar{p}_\kappa \simeq p_\kappa$ for the desired states:**
A good initial choice for $\bar{p}_\kappa$ is the corresponding root of the non-relativistic Sturmian secular equation (3.39).

(4) **Compute the relativistic T-matrix and diagonalize:**
Evaluate and diagonalize the matrix

$$T_{\nu',\nu} = Z \frac{\bar{p}_\kappa}{Q_\nu(\bar{p}_\kappa)} \delta_{\nu',\nu} + T'_{\nu',\nu}(\bar{p}_\kappa) \qquad (7.90)$$

using the method described in Appendix A. Because this matrix only weakly depends on $\bar{p}_\kappa$, the root closest to $\bar{p}_\kappa$ gives the relativistic value of the scaling parameter to a good approximation, and hence also the relativistic energy. If desired, step (4) can be repeated until a self-consistent relativistic root is obtained. In the authors' experience, this usually occurs after one or two repetitions.

In the above procedure, steps (1) and (2) can be done once and for all, yielding results that are valid for all states and nuclear charges, and the steps need not be repeated. Step (3) must be performed once for each nuclear charge, but potentially yields all the desired states at once. Step (4) must be performed both for each desired state and for each value of nuclear charge.

7.7 A simple example

When $j = \frac{1}{2}$, $l = 0$ and $M = \frac{1}{2}$, the relativistic hydrogenlike orbitals $\chi_{n,j,l,M}(\mathbf{x})$ have the form

$$\chi_\uparrow \equiv \chi_{n,\frac{1}{2},0,\frac{1}{2}}(\mathbf{x}) = \frac{1}{\sqrt{4\pi}} \begin{pmatrix} ig_\mu(r) \\ 0 \\ f_\mu(r)\cos\theta \\ f_\mu(r)\sin\theta e^{i\varphi} \end{pmatrix} \qquad (7.91)$$

while when $j=\frac{1}{2}$, $l=0$ and $M=-\frac{1}{2}$, they have the form

$$\chi_\downarrow \equiv \chi_{n,\frac{1}{2},0,-\frac{1}{2}}(\mathbf{x}) = \frac{1}{\sqrt{4\pi}} \begin{pmatrix} 0 \\ ig_\mu(r) \\ f_\mu(r)\sin\theta e^{-i\varphi} \\ -f_\mu(r)\cos\theta \end{pmatrix} \quad (7.92)$$

The corresponding densities are

$$(\chi_\uparrow^\dagger \chi_\uparrow) = \rho_{\uparrow,\uparrow} = \rho_{\downarrow,\downarrow} = \frac{1}{4\pi}[g_\mu(r)g_\mu(r) + f_\mu(r)f_\mu(r)] \quad (7.93)$$

and

$$\rho_{\uparrow,\downarrow} = 0 \quad (7.94)$$

The components of the spin-up transition current vector

$$\mathbf{j}_{\uparrow,\uparrow} = c(\chi_\uparrow^\dagger \boldsymbol{\alpha}\chi_\uparrow) \quad (7.95)$$

become

$$j_{x\ \uparrow,\uparrow} = -\frac{c}{2\pi}g_\mu(r)f_\mu(r)\sin\theta\sin\varphi$$

$$j_{y\ \uparrow,\uparrow} = \frac{c}{2\pi}g_\mu(r)f_\mu(r)\sin\theta\cos\varphi$$

$$j_{z\ \uparrow,\uparrow} = 0 \quad (7.96)$$

The current vectors corresponding respectively to the spin-up and spin-down states (7.91) and (7.92) are

$$\mathbf{j}_{\uparrow,\uparrow} = \frac{c}{2\pi}g_\mu(r)f_\mu(r)\sin\theta(-\sin\varphi,\cos\varphi,\ 0) = -\mathbf{j}_{\downarrow,\downarrow} \quad (7.97)$$

$\mathbf{j}_{\uparrow,\uparrow}$ and $\mathbf{j}_{\downarrow,\downarrow}$ can be thought of as small loops of current that generate vector potentials giving rise to magnetic fields associated with the electron's spin. These current-current interaction contributes to the Gaunt interaction shown in equation (7.77). For small values of Z, the contribution is extremely small. As Z increases, the contribution grows quadratically but still remains small compared with the density-density interaction.

In the previous section we discussed a four-step procedure for making relativistic calculation. As an illustration of this procedure, let us consider the ground state of atoms and ions in the heliumlike isoelectronic series using only the single Dirac configuration

$$\Phi_\nu = |\chi_{1,\frac{1}{2},0,\frac{1}{2}}\chi_{1,\frac{1}{2},0,-\frac{1}{2}}| \qquad (7.98)$$

The two Dirac spin-orbitals that appear in this configuration are shown in equations (7.91) and (7.92). We can now try to calculate the relativistic ground state energies by following the steps of page 94. Because we use only a single basis function, the results will of course be poor. However, our purpose here is not to obtain an accurate number, but to provide a simple illustration of the general procedure. An example using 16 configurations is shown in Figure 7.8.

Step 1: (*Compute Q_ν as a function of p_κ*)
We first tabulate (p_κ, Q_ν)-pairs for a range of Q_ν using equation (7.87):

$$\{(0,0), (14.1516, 10), (28.3603, 20), (42.6861, 30), (57.1947, 40)\}$$

For purposes of this simple example, we use only 5 points to compute an approximation to $Q_\nu(p_\kappa)$. The best cubic fit is

$$Q_\nu \simeq 0.707088 p_\kappa + 1.662427 \cdot 10^{-6} p_\kappa^2 - 2.389630 \cdot 10^{-6} p_\kappa^3 \qquad (7.99)$$

and is shown in Figure 7.4. Figure 7.5 illustrates the connection between $\frac{p_\kappa}{Q_\nu}$ in the relativistic case and $\mathcal{R}_\nu$ in the non-relativistic case.

Step 2: (*Compute $\mathcal{N}_\nu$ as a function of p_κ*)
In a manner similar to Step 1, we make a table of $(p_\kappa, \mathcal{N}_\nu)$ using equations (7.88) and (7.89):

$$\{(1, 1.), (10, 0.999668), (20, 0.998664), (30, 0.996982), (40, 0.994603)\}$$

The best quadratic fit is

$$\mathcal{N}_\nu \simeq 1 + 1.30372 \cdot 10^{-6} p_\kappa - 3.40379 \cdot 10^{-6} p_\kappa^2 \qquad (7.100)$$

Figure 7.6 shows the points together with the approximation.

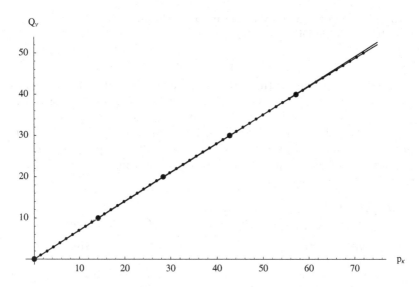

Fig. 7.4 (p_κ, Q_ν)-points plotted together with the best cubic fit for the points indicated by large dots, as well as a straight line included for comparison. The relationship is almost linear in this region, but not quite.

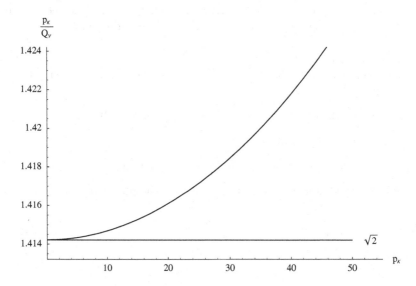

Fig. 7.5 This figure shows the ratio $\frac{p_\kappa}{Q_\nu}$ as a function of p_κ. In the relativistic Sturmian secular equation this ratio appears in the position occupied by $\mathcal{R}_\nu$ in the non-relativistic case. For the ground state of the heliumlike isoelectronic series, $\mathcal{R}_\nu = \sqrt{2}$. In the figure we see that $\frac{p_\kappa}{Q_\nu} \to \mathcal{R}_\nu$ as $p_\kappa \to 0$.

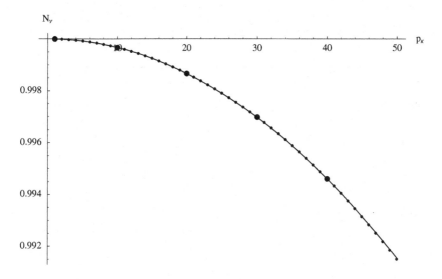

Fig. 7.6 $(p_\kappa, \mathcal{N}_\nu)$-points plotted together with the best quadratic fit for the points indicated by large dots. The intermediate points show that the relationship is very nearly quadratic.

Step 3: (*Obtain approximate root $\bar{p}_\kappa \simeq p_\kappa$*)

Because there is only a single basis function, the non-relativistic Sturmian secular equation (3.39) becomes simply

$$Z\mathcal{R}_\nu + T'^{(nr)}_{\nu\nu} - \bar{p}_\kappa = 0 \qquad (7.101)$$

By applying the methods of Appendix A, one obtains

$$T'^{(nr)}_{\nu\nu} = \frac{5}{8\sqrt{2}} \qquad (7.102)$$

Finally we have $\mathcal{R}_\nu = \sqrt{\frac{1}{n^2} + \frac{1}{n'^2}} = \sqrt{\frac{1}{1} + \frac{1}{1}} = \sqrt{2}$, and equation (7.101) becomes

$$\bar{p}_\kappa = Z\sqrt{2} + \frac{5}{8\sqrt{2}} \qquad (7.103)$$

Step 4: (*Compute relativistic T-matrix and diagonalize*)

Because there is only a single basis function, the relativistic Sturmian secular equation (7.85) becomes

$$Z\frac{p_\kappa}{Q_\nu} + T'_{\nu\nu} - p_\kappa = 0 \qquad (7.104)$$

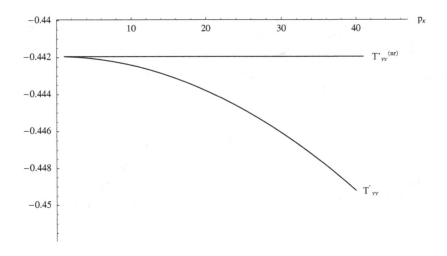

Fig. 7.7 This figure shows the relativistic $T'_{\nu\nu}$ as a function of p_κ (since there is only one basis function in our simple example, T' has only one element). For low values of p_κ it approaches the non-relativistic value, $T'^{(nr)}_{\nu\nu}$. The methods of Appendix A were used to calculate both $T'^{(nr)}_{\nu\nu}$ and $T'_{\nu\nu}$.

We compute $T'_{\nu\nu}$ as a function of p_κ using Q_ν and $\mathcal{N}_\nu$ from steps (1) and (2), and applying the methods discussed in Appendix A. The results are shown in Figure 7.7. Finally, iterating the relationship

$$\bar{p}_\kappa^{(i+1)} = Z \frac{\bar{p}_\kappa^{(i)}}{Q_\nu\left(\bar{p}_\kappa^{(i)}\right)} + T'_{\nu\nu}\left(\bar{p}_\kappa^{(i)}\right) \qquad (7.105)$$

starting with $\bar{p}_\kappa^{(0)} = \bar{p}_\kappa$ from step 3, we obtain the results shown in the following table:

	$Z = 2$	$Z = 10$	$Z = 20$	$Z = 30$	$Z = 50$
$\bar{p}_\kappa^{(0)} = Z\sqrt{2} + \frac{5}{8\sqrt{2}}$	2.38649	13.7002	27.8423	41.9845	70.2687
$\bar{p}_\kappa^{(1)}$	2.38658	13.7082	27.9120	42.2276	71.4366
$\bar{p}_\kappa^{(2)}$	2.38658	13.7082	27.9123	42.2305	71.4766
$\bar{p}_\kappa^{(3)}$	2.38658	13.7082	27.9123	42.2305	71.4780
$E_\kappa^{(nr)}$	−2.84766	−93.8477	−387.598	−881.348	−2468.8
E_κ	−2.84789	−93.9578	−389.549	−891.708	−2554.5
$E_\kappa^{(expt)}$	−2.90339	−94.0055	−389.489	−891.321	-

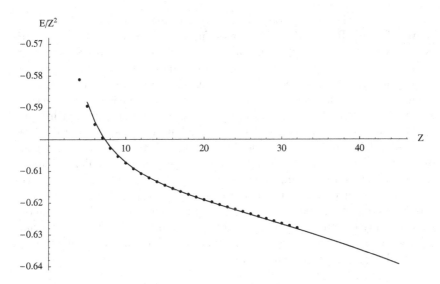

Fig. 7.8 This figure shows the calculated values of E/Z^2 as a function of Z for the (1s)(2s) ^{3}S states of the 2-electron isoelectronic series (smooth line) compared with experimental values taken from the NIST Tables (dots). For large values of Z, the relativistic calculation used here slightly overestimates the binding energies because of quantum electrodynamic effects. For small values of Z, the calculation slightly underestimates the binding energy because of basis set truncation.

7.8 Fine structure of spectral lines

In Chapter 3 we discussed the symmetry-adapted basis functions appropriate for building up the multiconfigurational non-relativistic wave function of an atom in the absence of external fields. We saw that these symmetry-adapted functions (called Russell-Saunders basis functions) are simultaneous eigenfunctions of the total orbital angular momentum operator $\mathbf{L}^2$, the z-component of total orbital momentum $\mathbf{L}_z$, the total spin operator $\mathbf{S}^2$, the z-component of total spin, $\mathbf{S}_z$, where

$$\mathbf{L} \equiv \mathbf{L}_1 + \mathbf{L}_2 + \cdots + \mathbf{L}_N$$
$$\mathbf{S} \equiv \mathbf{S}_1 + \mathbf{S}_2 + \cdots + \mathbf{S}_N \qquad (7.106)$$

The quantum numbers labeling the non-relativistic Russell-Saunders basis functions are L, M_L, S and M_S, and in the absence of spin-orbit coupling, states characterized by these quantum numbers are $(2L+1) \times (2S+1)$-fold degenerate.

In relativistic calculations, especially for high values of the nuclear charge Z, the Russell-Saunders basis functions are no longer appropriate as symmetry-adapted basis functions. This is because neither the z-component of the 1-electron orbital angular momentum operator nor the z-component of the 1-electron spin operator commutes with the Dirac Hamiltonian for an electron moving in the field of a nucleus. As we saw in equations (7.28-7.31), the 1-electron Dirac spinors $\chi_{n,j,l,M}(\mathbf{x})$ which represent an electron moving in the potential of a bare nucleus, are characterized by the quantum numbers j, l and M, while m_l and m_s are no longer good quantum numbers. Furthermore, when we replace the Coulomb interaction operator by the Gaunt interaction, magnetic interactions between the electrons are introduced. The result of both 1-electron and many-electron relativistic effects is that M_L and M_S can no longer be used to characterize the many-electron symmetry-adapted basis functions. The appropriate relativistic symmetry-adapted basis functions are instead eigenfunctions of the total angular momentum operator

$$\mathbf{J}^2 \equiv (\mathbf{L} + \mathbf{S})^2 \qquad (7.107)$$

and the operator corresponding to the z-component of total angular momentum,

$$\mathbf{J}_z \equiv (\mathbf{L} + \mathbf{S})_z \qquad (7.108)$$

where $\mathbf{L}$ and $\mathbf{S}$ are defined by (7.106). The 1-electron Dirac spinors (which are eigenfunctions of $(\mathbf{J}_1)^2$, $(\mathbf{J}_2)^2,\ldots,(\mathbf{J}_N)^2$ with quantum numbers $j_1, j_2, \ldots, j_N$, and simultaneously eigenfunctions of $(\mathbf{J}_1)_z$, $(\mathbf{J}_2)_z,\ldots,(\mathbf{J}_N)_z$, with quantum numbers $M_1, M_2,\ldots,M_N$) are combined to form eigenfunctions of the total angular momentum operator, $\mathbf{J}^2$, and its z-component, $\mathbf{J}_z$. In the Russell-Saunders scheme, one can also construct eigenfunctions of $\mathbf{J}^2$ and $\mathbf{J}_z$, but one begins with eigenfunctions of $\mathbf{L}^2$, $\mathbf{L}_z$, $\mathbf{S}^2$ and $\mathbf{S}_z$.

To illustrate the differences between non-relativistic and relativistic calculations and the changes in the relative magnitude of the various interactions as Z increases, we can consider the $n_1 = 1$, $n_2 = 2$ $\mathcal{R}$-block of the 2-electron isoelectronic series, i.e. the block of configurations for which

$$\mathcal{R}_\nu = \sqrt{\frac{1}{1^2} + \frac{1}{2^2}} \qquad (7.109)$$

Since $(2n_1^2) \times (2n_2^2) = 2 \times 8 = 16$, there are 16 primitive configurations belonging to this $\mathcal{R}$-block, and when Z is so large that interelectron repulsion can be neglected, these configurations are degenerate eigenfunctions of the

Table 7.2: Eigenvalues of $T'_{\nu',\nu}$ for the heliumlike $\mathcal{R}_\nu = \sqrt{5}/2$ block. When spin-orbit coupling is included in the non-relativistic calculation, the 9-fold degenerate ^{3}P level splits into a 5-fold degenerate level with $J=2$, a 3-fold degenerate level with $J=1$, and a non-degenerate level with $J=0$.

$|\lambda_\kappa|$	term	degen.	configuration	with spin-orbit coupling
0.168089	^{1}P	3	(1s)(2p)	^{1}P$_1$
0.201897	^{3}P	9	(1s)(2p)	^{3}P$_2$, degen.=5 ^{3}P$_1$, degen.=3 ^{3}P$_0$, degen.=1
0.207350	^{1}S	1	(1s)(2s)	^{1}S$_0$
0.232434	^{3}S	3	(1s)(2s)	^{3}S$_1$

non-relativistic Hamiltonian, corresponding to the energy

$$E_\kappa = -\frac{Z^2 \mathcal{R}_\nu^2}{2} = -\frac{Z^2}{2}\left(\frac{1}{1^2} + \frac{1}{2^2}\right) \tag{7.110}$$

When interelectron repulsion is added (still in the non-relativistic large-Z approximation, and still neglecting spin-orbit coupling), the 16-fold degeneracy is partly removed and we obtain a 3-fold degenerate ^{3}S level, a non-degenerate ^{1}S level, a 9-fold degenerate ^{3}P level, and a 3-fold degenerate ^{1}P level. In the non-relativistic large-Z approximation, the energies of these 4 levels are then given by

$$E_\kappa = -\frac{1}{2}(Z\mathcal{R}_\nu - |\lambda_\kappa|)^2 \tag{7.111}$$

where the λ_κ's are the roots of the non-relativistic interelectron repulsion matrix. When spin-orbit coupling is included in the non-relativistic calculation, the 9-fold degenerate ^{3}P level splits into a 5-fold degenerate level with $J=2$, a 3-fold degenerate level with $J=1$, and a non-degenerate level with $J=0$. The sum of the degeneracies is 16, as it should be.

Now let us consider what happens in a relativistic calculation when Z is so large that interelectron interactions can be neglected and where interactions with other $\mathcal{R}$-blocks are also neglected. From equation (7.40) we can see that the 1-electron energies will now depend on j as well as on n. The 16 primitive configurations will be characterized by the quantum numbers $n_1 = 1$, $l_1 = 0$, $j_1 = \frac{1}{2}$, $M_1 = -\frac{1}{2}, \frac{1}{2}$ for one of the Dirac spinors and $n_2 = 2$, $l_2 = 0, 1$, $j_2 = \frac{1}{2}$, $M_2 = -\frac{1}{2}, \frac{1}{2}$ or alternatively $n_2 = 2$, $l_2 = 1$, $j_2 = \frac{3}{2}$, $M_2 = -\frac{3}{2}, -\frac{1}{2}, \frac{1}{2}, \frac{3}{2}$ for the other. Since the relativistic 1-electron energies depend on j as well as n, the primitive configurations are no longer 16-fold degenerate when interelectron repulsion is completely neglected. We have instead two 8-fold degenerate levels with energies

$$E_{\frac{3}{2}} = -2c^2 + \epsilon_{1,\frac{1}{2}} + \epsilon_{2,\frac{3}{2}} \tag{7.112}$$

and

$$E_{\frac{1}{2}} = -2c^2 + \epsilon_{1,\frac{1}{2}} + \epsilon_{2,\frac{1}{2}} \tag{7.113}$$

where the 1-electron energies $\epsilon_{n,j}$ are given by equation (7.40).

When the Gaunt interaction is added, we obtain the calculated relativistic energies of the $\mathcal{R}$-block shown in Figure 7.5 as functions of Z. The calculation neglects interactions with other R-blocks, i.e. it is the relativistic analogue of the large-Z approximation discussed in Chapter 3 and shown in equation (7.111). As we reintroduce interelectron repulsion, the 8-fold degenerate level $E_{\frac{3}{2}}$ splits into a three-fold degenerate level characterized by the quantum numbers $J = 1$ and $M_J = -1, 0, 1$, and a 5-fold degenerate level with $J = 2$ and $M_J = -2, -1, 0, 1, 2$. The other 8-fold degenerate level, $E_{\frac{1}{2}}$, splits into two non-degenerate levels with $J = 0$ and $M_J = 0$, one with $l_2 = 0$ and the other with $l_2 = 1$, plus two 3-fold degenerate levels with $J = 1$ and $M_J = -1, 0, 1$, again one with $l_2 = 0$ and the other with $l_2 = 1$.

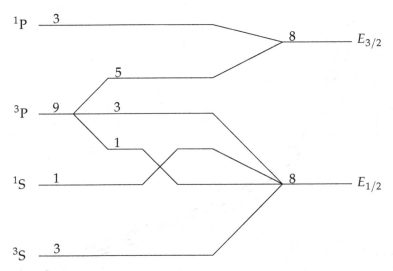

Fig. 7.9 Schematic figure of the fine structure of the $n = 1$, $n'=2$ levels for atoms and ions in the 2-electron isoelectronic series. On the left of the figure we see the levels in the non-relativistic approximation, neglecting spin-orbit coupling. When spin-orbit coupling is added, the ^{3}P level splits into the ^{3}P$_0$, ^{3}P$_1$ and ^{3}P$_2$ levels. On the extreme right of the figure we see the relativistic levels neglecting interelectron interactions. There are two 8-fold degenerate levels, whose energies are given by equations (7.112) and (7.113). When the Gaunt interaction is added, these 8-fold degenerate levels split and correlate with the non-relativistic levels in the manner shown in the figure. The numbers above the levels represent degeneracies.

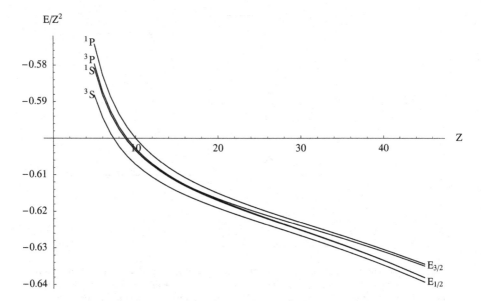

Fig. 7.10 Calculated fine structure for the $n = 1$, $n' = 2$ levels of the 2-electron isoelectronic series. For low values of Z, spin-orbit coupling is small, and the calculated multiplets correspond to those shown on the left-hand side of Figure 7.5. For large Z, the Gaunt interaction becomes less important compared with nuclear attraction, so that the energies approach $E_{3/2}$ and $E_{1/2}$.

Chapter 8

MOMENTUM SPACE; THE FOCK TRANSFORMATION

8.1 One-electron Coulomb Sturmians in direct space

In Chapter 1, we discussed one-electron Coulomb Sturmian basis sets. The functions in a basis set of this type have the form

$$\chi_{nlm}(\mathbf{x}) = R_{nl}(r)Y_{lm}(\theta,\phi) \qquad (8.1)$$

where the radial functions are defined by equations (2.2)-(2.4). The first few Coulomb Sturmian radial functions are

$$R_{10}(r) = 2k^{3/2}e^{-kr}$$

$$R_{20}(r) = 2k^{3/2}(1-kr)e^{-kr}$$

$$R_{21}(r) = \frac{2k^{3/2}}{\sqrt{3}} kr\, e^{-kr}$$

$$\vdots \quad \vdots \quad \vdots \qquad (8.2)$$

The Sturmian radial functions shown in equation (8.2) (and also in Table 1.1) are identical with the familiar hydrogenlike orbitals, except that Z/n in the hydrogenlike orbitals has been replaced by k, a constant that is the same for all the radial functions in the Sturmian basis set. Since the hydrogenlike orbitals obey a one-electron Schrödinger equation of the form

$$\left[-\frac{1}{2}\nabla^2 + \frac{1}{2}\frac{Z^2}{n^2} - \frac{Z}{r}\right]\chi'_{nlm}(\mathbf{x}) = 0 \qquad (8.3)$$

it follows, with the substitution $Z/n \to k$, that the Coulomb Sturmians obey

$$\left[-\frac{1}{2}\nabla^2 + \frac{1}{2}k^2 - \frac{nk}{r} \right] \chi_{nlm}(\mathbf{x}) = 0 \qquad (8.4)$$

In Chapter 1, we also showed that they obey a potential-weighted orthonormality relation of the form

$$\int d^3x \; \chi^*_{n'l'm'}(\mathbf{x}) \frac{1}{r} \chi_{nlm}(\mathbf{x}) = \frac{k}{n} \delta_{n'n} \delta_{l'l} \delta_{m'm} \qquad (8.5)$$

8.2 Fourier transforms of Coulomb Sturmians

A momentum-space basis set corresponding to the Coulomb Sturmians is defined by the relationships

$$\chi_{nlm}(\mathbf{x}) = \frac{1}{(2\pi)^{3/2}} \int d^3p \; e^{i\mathbf{p}\cdot\mathbf{x}} \chi^t_{nlm}(\mathbf{p})$$

$$\chi^t_{nlm}(\mathbf{p}) = \frac{1}{(2\pi)^{3/2}} \int d^3x \; e^{-i\mathbf{p}\cdot\mathbf{x}} \chi_{nlm}(\mathbf{x}) \qquad (8.6)$$

From the familiar expansion of a plane wave in terms of spherical harmonics and spherical Bessel functions

$$e^{-i\mathbf{p}\cdot\mathbf{x}} = 4\pi \sum_{l=0}^{\infty} (-i)^l j_l(pr) \sum_{m=-l}^{l} Y^*_{lm}(\theta, \phi) Y_{lm}(\theta_p, \phi_p) \qquad (8.7)$$

we can see that

$$\chi^t_{nlm}(\mathbf{p}) = R^t_{nl}(p) Y_{lm}(\theta_p, \phi_p) \qquad (8.8)$$

where

$$R^t_{nl}(p) = (-i)^l \sqrt{\frac{2}{\pi}} \int_0^\infty dr \; r^2 j_l(pr) R_{nl}(r) \qquad (8.9)$$

The integrals needed for the evaluation of $R^t_{nl}(p)$ have the form

$$J_{sl} \equiv \int_0^\infty dr \; r^s e^{-kr} j_l(pr) \qquad (8.10)$$

where j_l is a spherical Bessel function of order l. The most simple of these integrals is

$$J_{10} = \int_0^\infty dr \; r \; e^{-kr} j_0(pr) = \frac{1}{p} \int_0^\infty dr \; e^{-kr} \sin(pr) \qquad (8.11)$$

which can be evaluated by elementary methods, yielding

$$J_{10} = \frac{1}{p^2 + k^2} \tag{8.12}$$

The integrals J_{sl} corresponding to higher values of s and l can be found by differentiating with respect to k, e.g.

$$J_{20} = -\frac{\partial}{\partial k}\frac{1}{p^2 + k^2} = \frac{2k}{(p^2 + k^2)^2} \tag{8.13}$$

$$J_{30} = -\frac{\partial}{\partial k}\frac{2k}{(p^2 + k^2)^2} = \frac{2(3k^2 - p^2)}{(p^2 + k^2)^3} \tag{8.14}$$

and by means of the recursion relation [Harris and Michels, 1967]

$$kJ_{s,l} = pJ_{s,l-1} + (s - l - 1)J_{s-1,l} \tag{8.15}$$

Another useful recursion relation is

$$J_{l+1,l} = \frac{2lp}{p^2 + k^2} J_{l,l-1} \tag{8.16}$$

When applied repeatedly to J_{10}, this yields

$$J_{l+1,l} = \frac{2^l l! p^l}{(p^2 + k^2)^{l+1}} \tag{8.17}$$

The first few integrals J_{sl} are shown in Table 8.1, and others can be derived from them by the methods just discussed. Knowledge of these integrals allows us to evaluate $R_{nl}^t(p)$ in a completely straightforward way. For example, one finds in this way that

$$R_{10}^t(p) = \frac{1}{\sqrt{2\pi}} \frac{4k^{5/2}}{(p^2 + k^2)^2}$$

$$R_{20}^t(p) = \frac{-2}{\sqrt{2\pi}} \frac{4k^{5/2}}{(p^2 + k^2)^2} \frac{k^2 - p^2}{p^2 + k^2}$$

$$R_{21}^t(p) = \frac{-2i}{\sqrt{2\pi}} \frac{4k^{5/2}}{(p^2 + k^2)^2} \frac{kp}{p^2 + k^2}$$

$$\vdots \quad \vdots \quad \vdots \tag{8.18}$$

8.3 The Fock projection; Hyperspherical harmonics

In a remarkably brilliant early paper [Fock, 1935], [Fock, 1958], V. Fock demonstrated a relationship between Fourier transformed Sturmian basis sets and 4-dimensional hyperspherical harmonics. He introduced the following transformation

$$u_1 = \frac{2kp_1}{k^2 + p^2}$$

$$u_2 = \frac{2kp_2}{k^2 + p^2}$$

$$u_3 = \frac{2kp_3}{k^2 + p^2}$$

$$u_4 = \frac{k^2 - p^2}{k^2 + p^2} \qquad (8.19)$$

which projects 3-dimensional momentum space onto the surface of a 4-dimensional hypersphere. He was then able to show (see Appendix D) that

$$\chi_{n,l,m}^t(\mathbf{p}) = M(p) Y_{n-1,l,m}(\mathbf{u}) \qquad (8.20)$$

where

$$M(p) \equiv \frac{4k^{5/2}}{(k^2 + p^2)^2} \qquad (8.21)$$

is a universal function of p, independent of the quantum numbers n, l, and m. In equation (8.20), $Y_{\lambda,l,m}(\mathbf{u})$ is a 4-dimensional hyperspherical harmonic, the 4-dimensional generalization of the familiar spherical harmonics in 3 dimensions. The method by which Fock derived this remarkable result is discussed in Appendix D, while the general theory of hyperspherical harmonics will be developed in Chapters 9 and 10. The first few hyperspherical harmonics are given by

$$Y_{0,0,0}(\mathbf{u}) = \frac{1}{\sqrt{2}\pi}$$

$$Y_{1,0,0}(\mathbf{u}) = \frac{-2u_4}{\sqrt{2}\pi}$$

$$Y_{1,1,0}(\mathbf{u}) = \frac{-2iu_3}{\sqrt{2}\pi}$$

$$\vdots \quad \vdots \quad \vdots \qquad (8.22)$$

The reader may enjoy verifying that by combining equations (8.19)-(8.22), one obtains the Fourier-transformed Coulomb Sturmians shown in (8.18).

The number of linearly independent 4-dimensional hyperspherical harmonics corresponding to a particular value of the main index λ is $(\lambda+1)^2$, and Fock pointed out that with the identification $\lambda = n-1$, this corresponds to the degeneracy of the hydrogenlike orbitals, a degeneracy that in the conventional picture can only be explained by invoking dynamical symmetry.

The first few 4-dimensional hyperspherical harmonics are shown in Tables 8.2, 10.4 and 10.5. As can be seen from these tables, there are several alternative ways in which orthonormal sets of hyperspherical harmonics may be constructed. Those shown in Table 10.4 are the "standard set", which is based on the chain of subgroups $SO(4) \supset SO(3) \supset SO(2)$. The 4-dimensional hyperspherical harmonics of the standard set can be obtained from the formula

$$Y_{\lambda,l,m}(\mathbf{u}) = \mathcal{N}_{\lambda,l} C_{\lambda-l}^{1+l}(u_4) Y_{l,m}(u_1, u_2, u_3) \qquad (8.23)$$

where

$$\mathcal{N}_{\lambda,l} = (-1)^\lambda i^l (2l)!! \sqrt{\frac{2(\lambda+1)(\lambda-l)!}{\pi(\lambda+l+1)!}} \qquad (8.24)$$

is a normalizing constant

$$C_\lambda^\alpha(u_4) = \sum_{t=0}^{[\lambda/2]} \frac{(-1)^t \Gamma(\lambda+\alpha-t)}{t!(\lambda-2t)!\Gamma(\alpha)} (2u_4)^{\lambda-2t} \qquad (8.25)$$

is a Gegenbauer polynomial, and where $Y_{l,m}$ is a familiar 3-dimensional spherical harmonic.

8.4 The momentum-space orthonormality relations revisited

If the angles characterizing points on Fock's 4-dimensional hypersphere are defined by

$$u_1 = \frac{2kp_1}{k^2+p^2} = \sin\chi_p \sin\theta_p \cos\phi_p$$

$$u_2 = \frac{2kp_2}{k^2+p^2} = \sin\chi_p \sin\theta_p \sin\phi_p$$

$$u_3 = \frac{2kp_3}{k^2+p^2} = \sin\chi_p \cos\theta_p$$

$$u_4 = \frac{k^2 - p^2}{k^2 + p^2} = \cos\chi_p \qquad (8.26)$$

then with a certain amount of work, one can show that the 4-dimensional solid angle $d\Omega$ is related to the volume element in momentum space, d^3p by

$$d\Omega = \sin^2\chi_p \sin\theta_p \, d\chi_p d\theta_p d\phi_p = \left(\frac{2k}{k^2+p^2}\right)^3 d^3p \qquad (8.27)$$

Then, since the hyperspherical harmonics obey the orthonormality relation

$$\int d\Omega \, Y^*_{\lambda',l',m'} Y_{\lambda,l,m} = \delta_{\lambda',\lambda}\delta_{l',l}\delta_{m',m} \qquad (8.28)$$

it follows from equations (8.20), (8.21), (8.27) and (8.28) that the momentum-space Coulomb Sturmians obey the weighted orthonormality relation

$$\int d^3p \, \chi^{t*}_{n',l',m'}(\mathbf{p}) \left(\frac{k^2+p^2}{2k^2}\right) \chi^t_{n,l,m}(\mathbf{p}) = \delta_{n'n}\delta_{l'l}\delta_{m'm} \qquad (8.29)$$

Thus the orthonormality property of the hyperspherical harmonics gives us an alternative method for deriving the momentum-space orthonormality of Coulomb Sturmian basis sets. If we identify k with p_0, let $d=3$ and identify (n,l,m) with ν, equation (8.29) is the same as (2.23). The completeness property of the Coulomb Sturmians (which originally motivated their introduction into quantum theory), is related, through equation (8.20), to the completeness of the hyperspherical harmonics.

Because the set of Coulomb Sturmians is complete (in the sense that they span the Sobolev space $W_2^{(1)}(\mathbb{R}^3)$) we can try to use them to represent a plane wave as was done in Chapter 2. For $d=3$, the Sturmian expansion of a plane wave becomes

$$e^{i\mathbf{p}\cdot\mathbf{x}} = (2\pi)^{3/2} \left(\frac{k^2+p^2}{2k^2}\right) \sum_{nlm} \chi^{t*}_{nlm}(\mathbf{p})\chi_{nlm}(\mathbf{x}) \qquad (8.30)$$

which can alternatively be written in the form

$$e^{i\mathbf{p}\cdot\mathbf{x}} = (2\pi)^{3/2} \left(\frac{2k^{1/2}}{k^2+p^2}\right) \sum_{nlm} Y^*_{n-1,l,m}(\mathbf{u})\chi_{nlm}(\mathbf{x}) \qquad (8.31)$$

These expansions are not point-wise convergent, but are valid in the sense of distributions.

Table 8.1: $J_{sl} \equiv \int_0^\infty dr\, r^s e^{-kr} j_l(pr)$

l	J_{1l}	J_{2l}	J_{3l}	J_{4l}
0	$\dfrac{1}{p^2+k^2}$	$\dfrac{2k}{(p^2+k^2)^2}$	$\dfrac{2(3k^2-p^2)}{(p^2+k^2)^3}$	$\dfrac{24k(k^2-p^2)}{(p^2+k^2)^4}$
1		$\dfrac{2p}{(p^2+k^2)^2}$	$\dfrac{8pk}{(p^2+k^2)^3}$	$\dfrac{8p(5k^2-p^2)}{(p^2+k^2)^4}$
2			$\dfrac{8p^2}{(p^2+k^2)^3}$	$\dfrac{48p^2 k}{(p^2+k^2)^4}$
3				$\dfrac{48p^3}{(p^2+k^2)^4}$

Table 8.2: Hydrogenlike Sturmians and their Fourier transforms with $t_j \equiv kx_j$, $j = 1, 2, 3$, $t \equiv kr$ and with **u** defined by equation (8.19).

name	$\left(\dfrac{\pi}{k^3}\right)^{1/2} \chi_{n,l,m}(\mathbf{x})$	$\dfrac{\sqrt{2}\pi}{M(p)} \chi^t_{n,l,m}(\mathbf{p})$
1s	e^{-t}	1
2s	$e^{-t}(1-t)$	$-2u_4$
$2p_j$	$e^{-t} t_j$	$-2iu_j$
3s	$e^{-t}(1 - 2t + \dfrac{2t^2}{3})$	$4u_4^2 - 1$
$3p_j$	$\left(\dfrac{2}{3}\right)^{1/2} e^{-t}(2-t)t_j$	$2i\sqrt{6}\, u_4 u_j$

Table 8.3: Alternative 4-dimensional hyperspherical harmonics

τ	λ	l	$\sqrt{2}\pi\, Y_\tau(\mathbf{u})$
1s	0	0	1
$2p_1$	1	1	$-2iu_1$
$2p_2$	1	1	$-2iu_2$
$2p_3$	1	1	$-2iu_3$
2s	1	0	$-2u_4$
$3d_{z^2}$	2	2	$-\sqrt{2}(2u_3^2 - u_1^2 - u_2^2)$
$3d_{x^2-y^2}$	2	2	$-\sqrt{6}(u_1^2 - u_2^2)$
$3d_{xy}$	2	2	$-2\sqrt{6}\, u_1 u_2$
$3d_{yz}$	2	2	$-2\sqrt{6}\, u_2 u_3$
$3d_{zx}$	2	2	$-2\sqrt{6}\, u_3 u_1$
$3p_1$	2	1	$2i\sqrt{6}\, u_4 u_1$
$3p_2$	2	1	$2i\sqrt{6}\, u_4 u_2$
$3p_3$	2	1	$2i\sqrt{6}\, u_4 u_3$
$3s$	2	0	$4u_4^2 - 1$

Chapter 9

HARMONIC POLYNOMIALS

9.1 Monomials, homogeneous polynomials, and harmonic polynomials

In Chapter 8, we saw that by means of the Fock transformation, the Fourier transforms of a set of hydrogenlike Sturmian basis functions can be expressed in terms of hyperspherical harmonics. In this chapter, and in the following one, we will develop the mathematical theory of hyperspherical harmonics. This is best done by starting with the properties of harmonic polynomials. As we shall see, harmonic polynomials are very closely related to hyperspherical harmonics [Vilenkin, 1968], [Wen and Avery, 1985], [Avery, 1989], [Avery, 2000].

We can start our discussion by considering monomials of the form

$$m_n = x_1^{n_1} x_2^{n_2} x_3^{n_3} \cdots x_d^{n_d} \tag{9.1}$$

where

$$n_1 + n_2 + \cdots + n_d = n \tag{9.2}$$

The monomial shown in equation (9.1) is a product of the variables $x_1, x_2, \ldots, x_d$ which might, for example, represent Cartesian coordinates in a d-dimensional space, with each of the variables x_j raised to the power n_j. For the monomials that we will consider, n_j is restricted to be a positive integer or zero. The sum of these integers, n, is said to be the *degree* of the monomial. If we differentiate a monomial m_n with respect to one of the variables $x_1, x_2, \ldots, x_d$, we obtain:

$$\frac{\partial m_n}{\partial x_j} = n_j x_j^{-1} m_n \tag{9.3}$$

Then by summing equation (9.2) over j and using (9.3) we have

$$\sum_{j=1}^{d} x_j \frac{\partial m_n}{\partial x_j} = n m_n \qquad (9.4)$$

For example

$$m_4 = x_1^2 x_2 x_3 \qquad (9.5)$$

is called a monomial of degree 4 because the sum of the integral powers to which the variables are raised is equal to 4:

$$n_1 + n_2 + n_3 = 2 + 1 + 1 = 4 \qquad (9.6)$$

We can check that equation (9.4) holds in this example, since

$$\sum_{j=1}^{d} x_j \frac{\partial m_4}{\partial x_j} = x_1(2x_1 x_2 x_3) + x_2(x_1^2 x_3) + x_3(x_1^2 x_2) = 4 x_1^2 x_2 x_3 = 4 m_4 \qquad (9.7)$$

A *homogeneous* polynomial of degree n is defined to be a linear combination of monomials, all of which are of degree n. We will use the symbol f_n to denote such a homogeneous polynomial. For example

$$f_4 = 2 x_1^2 x_2 x_3 + 3 x_4^4 + 4 x_5^2 x_6^2 \qquad (9.8)$$

is a homogeneous polynomial of degree 4 because it is composed of monomials all of which have this degree. From equation (9.4) and from the definition of homogeneous polynomials it follows that

$$\sum_{j=1}^{d} x_j \frac{\partial f_n}{\partial x_j} = n f_n \qquad (9.9)$$

Although it seems very elementary and obvious, equation (9.9) is an important relationship: It is the foundation upon which much of the theory of harmonic polynomials and hyperspherical harmonics can be built.

In order to explore the consequences of equation (9.9), we introduce the generalized Laplacian operator

$$\Delta \equiv \sum_{j=1}^{d} \frac{\partial^2}{\partial x_j^2} \qquad (9.10)$$

and the hyperradius

$$r^2 \equiv \sum_{j=1}^{d} x_j^2 \qquad (9.11)$$

Then, applying the generalized Laplacian operator to the product $r^\beta f_\alpha$ (where α and β are integers), and making use of (9.9), we obtain:

$$\Delta \left(r^\beta f_\alpha \right) = \beta(\beta + d + 2\alpha - 2) r^{\beta-2} f_\alpha + r^\beta \Delta f_\alpha \qquad (9.12)$$

We next define a *harmonic polynomial* h_n to be a homogeneous polynomial that is also a solution of the generalized Laplace equation:

$$\Delta h_n = 0 \qquad (9.13)$$

Here n denotes the harmonic polynomial's degree. For example,

$$h_4 = 3x_4^2 x_5 x_6 - x_5^3 x_6 \qquad (9.14)$$

is a harmonic polynomial of degree 4. It is homogeneous, but it is also a solution to (9.13). The reader may enjoy acting upon this polynomial with the generalized Laplacian operator defined in equation (9.10) to confirm that the result is zero.

Combining (9.12) and (9.13), we obtain:

$$\Delta \left(r^\beta h_\alpha \right) = \beta(\beta + d + 2\alpha - 2) r^{\beta-2} h_\alpha \qquad (9.15)$$

The final term on the right-hand side of equation (9.12) of course disappears because besides being homogeneous, h_n is also harmonic.

9.2 The canonical decomposition of a homogeneous polynomial

We shall next use equation (9.15) to show how every homogeneous polynomial f_n can be decomposed into a sum of harmonic polynomials multiplied by powers of the hyperradius:

$$f_n = h_n + r^2 h_{n-2} + r^4 h_{n-4} + \cdots \qquad (9.16)$$

This is called the *canonical decomposition* of the homogeneous polynomial. To see how this decomposition may be accomplished, we begin by acting on both sides of (9.16) with successively higher powers of Δ, making use of equation (9.15):

$$\Delta f_n = 2(d + 2n - 4) h_{n-2} + 4(d + 2n - 6) r^2 h_{n-4} + \cdots$$

$$\Delta^2 f_n = 8(d + 2n - 6)(d + 2n - 8) h_{n-4} + \cdots$$

$$\Delta^3 f_n = 48(d - 2n - 8)(d - 2n - 10)(d - 2n - 12) h_{n-6} + \cdots \qquad (9.17)$$

and so on. In general we obtain:

$$\Delta^\nu f_n = \sum_{k=\nu}^{\lfloor \frac{n}{2} \rfloor} \frac{(2k)!!}{(2k-2\nu)!!} \frac{(d+2n-2k-2)!!}{(d+2n-2k-2\nu-2)!!} r^{2k-2\nu} h_{n-2k} \qquad (9.18)$$

where

$$j!! \equiv \begin{cases} j(j-2)(j-4)\cdots 4 \times 2 & j = \text{even} \\ j(j-2)(j-4)\cdots 3 \times 1 & j = \text{odd} \end{cases} \qquad (9.19)$$

The upper limit in the sum shown in equation (9.18) is the "floor" $\lfloor \frac{n}{2} \rfloor$, which is the largest integer less than or equal to $\frac{n}{2}$. The set of simultaneous equations (9.18) can be solved to give expressions for the harmonic polynomials in the canonical decomposition of f_n.

As a simple example of equations (9.16)-(9.19), we can consider the canonical decomposition of a homogeneous polynomial of degree 2. Then (9.16) becomes

$$f_2 = h_2 + r^2 h_0 \qquad (9.20)$$

while from (9.17) or (9.18) we have

$$\Delta f_2 = \Delta(r^2 h_0) = 2d h_0 \qquad (9.21)$$

Equations (9.20) and (9.21) yield

$$\begin{aligned} h_0 &= \frac{1}{2d} \Delta f_2 \\ h_2 &= f_2 - \frac{r^2}{2d} \Delta f_2 \end{aligned} \qquad (9.22)$$

These are the harmonic polynomials that occur in the canonical decomposition of f_2. It is possible to use the same method for other degrees of f_n.

An alternative approach is to assume that the harmonic polynomial of highest degree in (9.16) can be written in the form

$$h_n = f_n + a_2 r^2 \Delta f_n + a_4 r^4 \Delta^2 f_n + a_6 r^6 \Delta^3 f_n + \cdots \qquad (9.23)$$

We can then try to find expressions for the coefficients $a_2, a_4, a_6, \ldots$. Applying Δ to both sides of (9.24), we have

$$0 = \Delta f_n + a_2 \Delta(r^2 \Delta f_n) + a_4 \Delta(r^4 \Delta^2 f_n) + \cdots \qquad (9.24)$$

We know that Δf_n is a homogeneous polynomial of degree $n - 2$, $\Delta^2 f_n$ is a homogeneous polynomial of degree $n - 4$, and so on. Making use of this knowledge, and also making use of (9.12), we obtain the relationships

$$1 + 2(d + 2n - 4)a_2 = 0$$
$$a_2 + 4(d + 2n - 6)a_4 = 0$$
$$a_4 + 6(d + 2n - 8)a_6 = 0$$
$$\vdots \quad \vdots \tag{9.25}$$

which can be solved to yield $a_2, a_4, a_6, \ldots$. This gives us a general expression for the harmonic polynomial of highest degree in the canonical decomposition:

$$h_n = f_n - \frac{r^2}{2(d + 2n - 4)} \Delta f_n + \frac{r^4}{8(d + 2n - 4)(d + 2n - 6)} \Delta^2 f_n + \cdots$$
$$= \sum_{j=0}^{\lfloor n/2 \rfloor} \frac{(-1)^j (d + 2n - 2j - 4)!!}{(2j)!!(d + 2n - 4)!!} r^{2j} \Delta^j f_n \tag{9.26}$$

Avery and Wen [Avery, 1989] have shown that the remaining harmonic polynomials in the decomposition are given by

$$h_{n-2\nu} = \frac{(d + 2n - 4\nu - 2)!!}{(2\nu)!!(d + 2n - 2\nu - 2)!!}$$
$$\times \sum_{j=0}^{\lfloor \frac{n}{2} - \nu \rfloor} \frac{(-1)^j (d + 2n - 4\nu - 2j - 4)!!}{(2j)!!(d + 2n - 4\nu - 4)!!} r^{2j} \Delta^{j+\nu} f_n \tag{9.27}$$

The logic behind (9.27) is as follows: $\Delta^\nu f_n$ is a homogeneous polynomial of degree $n - 2\nu$, and therefore its canonical decomposition can be written in the form

$$\Delta^\nu f_n = h'_{n-2\nu} + r^2 h'_{n-4\nu} + r^4 h'_{n-6\nu} + \cdots \tag{9.28}$$

From (9.26), we have

$$h'_{n-2\nu} = \sum_{j=0}^{\lfloor \frac{n}{2} - \nu \rfloor} \frac{(-1)^j (d + 2n - 4\nu - 2j - 4)!!}{(2j)!!(d + 2n - 4\nu - 4)!!} r^{2j} \Delta^{j+\nu} f_n \tag{9.29}$$

On the other hand, we know from equation (9.18) that

$$\Delta^\nu f_n = \frac{(2\nu)!!(d + 2n - 2\nu - 2)!!}{(d + 2n - 4\nu - 2)!!} h_{n-2\nu} + \cdots \tag{9.30}$$

Comparing (9.28) and (9.30), we can see that

$$h_{n-2\nu} = \frac{(d+2n-4\nu-2)!!}{(2\nu)!!(d+2n-2\nu-2)!!} h'_{n-2\nu} \qquad (9.31)$$

and this equation can be combined with (9.29) to yield (9.27).
Equation (9.27) can alternatively be written in the form

$$h_\lambda = \frac{(d+2\lambda-2)!!}{(n-\lambda)!!(d+n+\lambda-2)!!}$$
$$\times \sum_{j=0}^{\lfloor \lambda/2 \rfloor} \frac{(-1)^j (d+2\lambda-2j-4)!!}{(2j)!!(d+2\lambda-4)!!} r^{2j} \Delta^{j+\frac{1}{2}(n-\lambda)} f_n \qquad (9.32)$$

where we have replaced $n - 2\nu$ by λ. A particularly important special case of (9.32) is the one where $\lambda = 0$:

$$h_0 = \frac{(d-2)!!}{n!!(d+n-2)!!} \Delta^{\frac{1}{2}n} f_n \qquad (9.33)$$

The reader may enjoy using (9.32) to find the harmonic polynomials in the canonical decomposition of f_4:

$$f_4 = h_4 + r^2 h_2 + r^4 h_0 \qquad (9.34)$$

The result should be

$$h_0 = \frac{1}{8d(d+2)} \Delta^2 f_4$$

$$h_2 = \frac{1}{2(d+4)} \Delta f_4 - \frac{r^2}{4d(d+4)} \Delta^2 f_4$$

$$h_4 = f_4 - \frac{r^2}{2(d+4)} \Delta f_4 + \frac{r^4}{8(d+2)(d+4)} \Delta^2 f_4 \qquad (9.35)$$

9.3 Generalized angular momentum

We mentioned above that harmonic polynomials are closely related to hyperspherical harmonics. In discussing the relationship between them, it is useful to introduce the generalized angular momentum operator Λ^2, defined by

$$\Lambda^2 \equiv -\sum_{i>j}^d \sum_{j=1}^d \left(x_i \frac{\partial}{\partial x_j} - x_j \frac{\partial}{\partial x_i} \right)^2 \qquad (9.36)$$

When $d = 3$, Λ^2 reduces to the familiar orbital angular momentum operator

$$L^2 = L_1^2 + L_2^2 + L_3^2 \tag{9.37}$$

where

$$L_1 = \frac{1}{i}\left(x_2\frac{\partial}{\partial x_3} - x_3\frac{\partial}{\partial x_2}\right) \tag{9.38}$$

and where L_2 and L_3 are found by cyclic permutation of the indices in equation (9.38). Equation (9.36) can be rewritten in the form

$$\Lambda^2 = -r^2\Delta + \sum_{i,j=1}^{d} x_i x_j \frac{\partial^2}{\partial x_i \partial x_j} + (d-1)\sum_{i=1}^{d} x_i \frac{\partial}{\partial x_i} \tag{9.39}$$

It follows that

$$\Lambda^2 f_n = -r^2\Delta f_n + n(d-1)f_n + \sum_{i,j=1}^{d} x_i x_j \frac{\partial^2 f_n}{\partial x_i \partial x_j} \tag{9.40}$$

where we have made use of equation (9.9). In a manner similar to that used in establishing equation (9.9), one can show that

$$\sum_{i,j=1}^{d} x_i x_j \frac{\partial^2 f_n}{\partial x_i \partial x_j} = n(n-1)f_n \tag{9.41}$$

Finally, substituting (9.41) into (9.40), we obtain

$$\Lambda^2 f_n = -r^2\Delta f_n + n(n+d-2)f_n \tag{9.42}$$

In the case of harmonic polynomials this reduces to

$$\Lambda^2 h_\lambda = \lambda(\lambda + d - 2)h_\lambda \tag{9.43}$$

and for the special case where $d = 3$, we have

$$L^2 h_l = l(l+1)h_l \tag{9.44}$$

Thus harmonic polynomials can be seen to be eigenfunctions of the generalized angular momentum operator, and when $d = 3$, they are eigenfunctions of the familiar angular momentum operator L^2. The canonical decomposition of a homogeneous polynomial can therefore be seen as a decomposition into generalized angular momentum eigenfunctions.

9.4 Hyperangular integration

The properties of harmonic polynomials can be used to perform hyperangular integrations [Avery, 1998]. In a 3-dimensional space, the familiar relationship between the volume element and the solid angle element $d\Omega$ has the form

$$dx_1 dx_2 dx_3 = r^2 dr\, d\Omega \qquad (9.45)$$

In a similar way, the solid angle element $d\Omega$ in a d-dimensional space can be defined by the relationship:

$$dx_1 dx_2 \cdots dx_d = r^{d-1} dr\, d\Omega \qquad (9.46)$$

where r is the hyperradius of equation (9.11). Just as the integral over solid angle of the product of two eigenfunctions of angular momentum vanishes when they do not correspond to the same quantum number l (a relationship that can be established using the hermiticity of L^2), it follows from the hermiticity of Λ^2 that

$$\int d\Omega\, h_\lambda^* h_{\lambda'} = 0 \quad \text{if } \lambda \neq \lambda' \qquad (9.47)$$

For the particular case where $\lambda' = 0$, we have

$$\int d\Omega\, h_\lambda^* h_0 = h_0 \int d\Omega\, h_\lambda^* = 0 \quad \text{if } \lambda \neq 0 \qquad (9.48)$$

so that

$$\int d\Omega\, h_\lambda = 0 \quad \text{if } \lambda \neq 0 \qquad (9.49)$$

If we combine equation (9.49) with the canonical decomposition of a homogeneous polynomial, (9.16), we can see that

$$\int d\Omega\, f_n = \begin{cases} r^n h_0 \int d\Omega & n = \text{even} \\ 0 & n = \text{odd} \end{cases} \qquad (9.50)$$

where f_n is any homogeneous polynomial of degree n, and h_0 is the constant term that occurs in its canonical decomposition if n is even. If n is odd, the constant $\lambda = 0$ term does not occur in the canonical decomposition (9.16), and the result of the hyperangular integration is zero. To complete the evaluation of the integral (9.50), we need to evaluate the total solid angle $\int d\Omega$ in a d-dimensional space. This can be done by considering the volume integral of e^{-r^2}:

$$\int_0^\infty dr\, r^{d-1} e^{-r^2} \int d\Omega = \prod_{j=1}^d \int_{-\infty}^\infty dx_j\, e^{-x_j^2} \qquad (9.51)$$

On the left-hand side of (9.51), the integral is expressed in terms of the hyperradius and the generalized solid angle element, while on the right-hand side it is expressed in terms of Cartesian coordinates. Both the hyperradial integral on the left and the integrals involving Cartesian coordinates can be evaluated in terms of gamma functions:

$$\int_0^\infty dr\, r^{d-1} e^{-r^2} = \frac{\Gamma(d/2)}{2} \qquad (9.52)$$

while

$$\int_{-\infty}^\infty dx_j\, e^{-x_j^2} = \Gamma(1/2) = \pi^{\frac{1}{2}} \qquad (9.53)$$

Combining equations (9.51)-(9.53), we have

$$\int d\Omega = \frac{2\pi^{\frac{d}{2}}}{\Gamma\left(\frac{d}{2}\right)} \qquad (9.54)$$

Finally, with the help of equations (9.33) and (9.54), our hyperangular integration formula (9.50) becomes

$$\int d\Omega\, f_n = \begin{cases} \dfrac{2\pi^{d/2} r^n (d-2)!!}{\Gamma(d/2) n!! (d+n-2)!!} \Delta^{\frac{1}{2}n} f_n & n = \text{even} \\ 0 & n = \text{odd} \end{cases} \qquad (9.55)$$

This holds for any homogeneous polynomial f_n. Now let us consider some function of the variables $x_1, x_2, \ldots, x_d$ that may be expanded in a convergent series of the form

$$F(\mathbf{x}) = \sum_{n=0}^\infty f_n(\mathbf{x}) \qquad (9.56)$$

Then from (9.55) we have

$$\int d\Omega\, F(\mathbf{x}) = \frac{(d-2)!! 2\pi^{d/2}}{\Gamma\left(\frac{d}{2}\right)} \sum_{n=0,2,\ldots}^\infty \frac{r^n}{n!!(n+d-2)!!} \Delta^{n/2} f_n(\mathbf{x}) \qquad (9.57)$$

Making use of the fact that

$$\left\lfloor \Delta^{n/2} F(\mathbf{x}) \right\rfloor_{\mathbf{x}=0} = \Delta^{n/2} f_n(\mathbf{x}) \qquad (9.58)$$

and letting $n = 2\nu$, we can rewrite (9.57) in the form [Avery, 2000]:

$$\int d\Omega \ F(\mathbf{x}) = \frac{(d-2)!!2\pi^{d/2}}{\Gamma\left(\frac{d}{2}\right)} \sum_{\nu=0}^{\infty} \frac{r^{2\nu}}{(2\nu)!!(d+2\nu-2)!!} \left\lfloor \Delta^{\nu} F(\mathbf{x}) \right\rfloor_{\mathbf{x}=0} \qquad (9.59)$$

In the special case where $d = 3$, this reduces to

$$\int d\Omega \ F(\mathbf{x}) = 4\pi \sum_{\nu=0}^{\infty} \frac{r^{2\nu}}{(2\nu+1)!} \left\lfloor \Delta^{\nu} F(\mathbf{x}) \right\rfloor_{\mathbf{x}=0} \qquad (9.60)$$

Equations (9.59) and (9.60) hold for any function whatever, provided that it can be expended in a convergent series of polynomials. Notice that to use these formulas, one does not need to actually make the expansion, but such an expansion must be possible. The remarkable feature of the angular integration formulas (9.55), (9.59) and (9.60) is that integration has been replaced by differentiation. Since programs like Mathematica and Maple are able to perform differentiations very rapidly, these formulas are extremely powerful and convenient.

Table 9.1: Decomposition of monomials into harmonic polynomials. The indices i, j, k are assumed to be all unequal. The terms in brackets are harmonic.

n	$m_n = h_n + r^2 h_{n-2} + \cdots$
1	$x_i = (x_i)$
2	$x_i^2 = \left(x_i^2 - \dfrac{r^2}{d}\right) + r^2\left(\dfrac{1}{d}\right)$ $x_i x_j = (x_i x_j)$
3	$x_i^3 = \left(x_i^3 - \dfrac{3r^2 x_i}{d+2}\right) + r^2\left(\dfrac{3x_i}{d+2}\right)$ $x_i^2 x_j = \left(x_i^2 x_j - \dfrac{r^2 x_j}{d+2}\right) + r^2\left(\dfrac{x_j}{d+2}\right)$ $x_i x_j x_k = (x_i x_j x_k)$

Table 9.2: Canonical decomposition of homogeneous polynomials.

n	$f_n = h_n + r^2 h_{n-2} + r^4 h_{n-4} + \cdots$
1	$h_1 = f_1$
2	$h_2 = f_2 - \dfrac{r^2}{2d}\Delta f_2$ $h_0 = \dfrac{r^2}{2d}\Delta f_2$
3	$h_3 = f_3 - \dfrac{r^2}{2(d+2)}\Delta f_3$ $h_1 = \dfrac{r^2}{2(d+2)}\Delta f_3$
4	$h_4 = f_4 - \dfrac{r^2}{2(d+4)}\Delta f_4 + \dfrac{r^4}{8(d+2)(d+4)}\Delta^2 f_4$ $h_2 = \dfrac{1}{2(d+4)}\Delta f_4 - \dfrac{r^2}{4d(d+4)}\Delta^2 f_4$ $h_0 = \dfrac{1}{8d(d+2)}\Delta^2 f_4$

Chapter 10

HYPERSPHERICAL HARMONICS

10.1 The relationship between harmonic polynomials and hyperspherical harmonics

We can define hyperspherical harmonics, $Y_\lambda(\mathbf{u})$, by the relationship

$$Y_\lambda(\mathbf{u}) \equiv r^{-\lambda} h_\lambda(\mathbf{x}) \tag{10.1}$$

where r is the hyperradius, $h_\lambda(\mathbf{x})$ is a harmonic polynomial of degree λ, and where $\mathbf{u}$ is the d-dimensional unit vector

$$\mathbf{u} \equiv \frac{1}{r}(x_1, x_2, \ldots, x_d) = (u_1, u_2, \ldots, u_d) \tag{10.2}$$

In Chapter 9, we derived the relationship

$$\Lambda^2 h_\lambda = \lambda(\lambda + d - 2) h_\lambda \tag{10.3}$$

Since $h_\lambda = r^\lambda Y_\lambda(\mathbf{u})$, and since the generalized angular momentum operator commutes with the hyperradius, equation (10.3) implies that

$$\Lambda^2 Y_\lambda(\mathbf{u}) = \lambda(\lambda + d - 2) Y_\lambda(\mathbf{u}) \tag{10.4}$$

Equation (10.4) states that a hyperspherical harmonic of degree λ is an eigenfunction of generalized angular momentum corresponding to the eigenvalue $\lambda(\lambda + d - 2)$. When $d = 3$, this reduces to the familiar relationship

$$L^2 Y_l(\mathbf{u}) = l(l+1) Y_l(\mathbf{u}) \tag{10.5}$$

The reader might object that something is wrong with equation (10.5). In order to be *really* familiar, it ought to be

$$L^2 Y_{l,m}(\mathbf{u}) = l(l+1) Y_{l,m}(\mathbf{u}) \tag{10.6}$$

The additional index m indicates, of course, that besides being an eigenfunction of L^2, $Y_{l,m}(\mathbf{u})$ is also an eigenfunction of L_3, the z-component of orbital angular momentum, i.e.

$$L_3 Y_{l,m}(\mathbf{u}) = m Y_{l,m}(\mathbf{u}) \tag{10.7}$$

where

$$L_3 \equiv \frac{1}{i}\left(x_1 \frac{\partial}{\partial x_2} - x_2 \frac{\partial}{\partial x_1}\right) \tag{10.8}$$

In the language of group theory, one can say that the set of spherical harmonics $Y_{l,m}(\mathbf{u})$ form a basis for irreducible representations of the continuous rotation group $SO(2)$ in the 2-dimensional space of x_1 and x_2. The spherical harmonics also form a basis for irreducible representations of $SO(3)$, of which $SO(2)$ is a subgroup: $SO(3) \supset SO(2)$. Similarly, in a general d-dimensional space, one can use the chain of subgroups [Vilenkin, 1968]

$$SO(d) \supset SO(d-1) \supset SO(d-2) \supset \cdots \supset SO(2) \tag{10.9}$$

to organize a set of $d-2$ additional indices that distinguish between the various linearly independent hyperspherical harmonics all of which correspond to a particular eigenvalue of the generalized angular momentum operator Λ^2. If we use the symbol $\mu \equiv (\mu_1, \mu_2, \ldots, \mu_{d-2})$ to stand for all of these additional indices, then we can write

$$\Lambda_d^2 Y_{\lambda,\mu}(\mathbf{u}) = \lambda(\lambda+d-2) Y_{\lambda,\mu}(\mathbf{u})$$

$$\Lambda_{d-1}^2 Y_{\lambda,\mu}(\mathbf{u}) = \mu_1(\mu_1+d-3) Y_{\lambda,\mu}(\mathbf{u})$$

$$\Lambda_{d-2}^2 Y_{\lambda,\mu}(\mathbf{u}) = \mu_2(\mu_2+d-4) Y_{\lambda,\mu}(\mathbf{u})$$

$$\vdots \qquad \vdots \qquad \vdots$$

$$\tag{10.10}$$

where we have added a subscript d to the generalized angular momentum operator to indicate the dimension of the space within which it acts.

$$\Lambda_d^2 \equiv -\sum_{i>j}^{d}\sum_{j=1}^{d}\left(x_i \frac{\partial}{\partial x_j} - x_j \frac{\partial}{\partial x_i}\right)^2 \tag{10.11}$$

When the functions are properly normalized, a set of hyperspherical harmonics that obeys the equations (10.10) also obeys the orthonormality relations

$$\int d\Omega \, Y^*_{\lambda',\mu'}(\mathbf{u}) Y_{\lambda,\mu}(\mathbf{u}) = \delta_{\lambda',\lambda}\delta_{\mu',\mu} \qquad (10.12)$$

a relationship that can be established by using the hermiticity properties of the generalized angular momentum operators. In equation (10.12),

$$\delta_{\mu',\mu} \equiv \delta_{\mu'_1,\mu_1}\delta_{\mu'_2,\mu_2}\delta_{\mu'_3,\mu_3}\cdots \qquad (10.13)$$

The indices μ need not be organized according to the chain of subgroups shown in equation (10.9). Aquilanti and his co-workers [Aquilanti et al., 1995], [Aquilanti et al., 1996], [Coletti, 1998], [Caligiana, 2003], as well as a number of Russian authors, have studied sets of hyperspherical harmonics corresponding to alternative chains of subgroups, and these alternative hyperspherical harmonics may be more convenient than the standard set, depending on the physical problem being studied.

10.2 Construction of hyperspherical harmonics by means of harmonic projection

Let $h_\mu(x_1,\ldots,x_{d-1})$ be a harmonic polynomial of degree μ_1 in a $(d-1)$-dimensional space. Then

$$f_{\lambda,\mu}(x_1,\ldots,x_d) = x_d^{\lambda-\mu_1} h_\mu(x_1,\ldots,x_{d-1}) \qquad (10.14)$$

will be a homogeneous polynomial of degree λ, and (from equation (9.26))

$$O_\lambda[f_{\lambda,\mu}] = \sum_{j=0}^{[\lambda/2]} \frac{(-1)^j (d+2\lambda-2j-4)!!}{(2j)!!(d+2\lambda-4)!!} r^{2j} \Delta^j f_{\lambda,\mu} \qquad (10.15)$$

will be the harmonic polynomial of highest degree in its canonical decomposition. In equation (10.15), O_λ can be thought of as a projection operator that projects out a harmonic polynomial of degree λ from a function on which it acts. Since

$$\Delta h_\mu(x_1,\ldots,x_{d-1}) = 0 \qquad (10.16)$$

it follows that

$$\Delta^j [x_d^{\lambda-\mu_1} h_\mu(x_1,\ldots,x_{d-1})] = \frac{(\lambda-\mu_1)!}{(\lambda-\mu_1-2j)!} x_d^{\lambda-\mu_1-2j} h_\mu(x_1,\ldots,x_{d-1}) \qquad (10.17)$$

Making use of equations (10.14) and (10.17), we can rewrite the harmonic projection shown in (10.15) in the form

$$h_{\lambda,\mu}(x_1, x_2, \ldots, x_d) = O_\lambda \left[x_d^{\lambda - \mu_1} h_\mu(x_1, \ldots, x_{d-1}) \right]$$
$$= h_\mu(x_1, \ldots, x_{d-1}) \frac{(\lambda - \mu_1)!}{(d + 2\lambda - 4)!!}$$
$$\times \sum_{j=0}^{[(\lambda - \mu_1)/2]} \frac{(-1)^j (d + 2\lambda - 2j - 4)!!}{(2j)!!(\lambda - \mu_1 - 2j)!} r^{2j} x_d^{\lambda - \mu_1 - 2j} \quad (10.18)$$

Equation (10.18) can be used to construct sets of hyperspherical harmonics that are simultaneous eigenfunctions of $\Lambda_d^2, \Lambda_{d-1}^2, \Lambda_{d-2}^2, \ldots$, like those shown in equation (10.10). In the next section, we will illustrate this method by using it to construct the familiar 3-dimensional spherical harmonics as well as hyperspherical harmonics in a 4-dimensional space. Table 1.1 shows equation (10.18) for various values of $\lambda - \mu_1$.

10.3 Hyperspherical harmonics in a 4-dimensional space

To see how equation (10.18) can be used to construct spherical harmonics and hyperspherical harmonics, we can begin by constructing a set of harmonic polynomials in the 2-dimensional space of the coordinates x_1 and x_2. Since

$$\frac{\partial^2}{\partial x_1^2}(x_1 \pm ix_2)^m = m(m-1)(x_1 \pm ix_2)^m \quad (10.19)$$

and

$$\frac{\partial^2}{\partial x_2^2}(x_1 \pm ix_2)^m = -m(m-1)(x_1 \pm ix_2)^m \quad (10.20)$$

Adding these two equations yields

$$\Delta(x_1 \pm ix_2)^m = 0 \quad (10.21)$$

Thus

$$h_m(x_1, x_2) = (x_1 \pm ix_2)^m \quad (10.22)$$

is a harmonic polynomial of degree m in the 2-dimensional space. For each value of m there are 2 linearly independent harmonic polynomials of this type, one corresponding to the + sign in equation (10.22), and the other corresponding to the − sign.

When $d = 3$, $\lambda = l$, $\mu_1 = m$, and $h_\mu(x_1, x_2) = (x_1 \pm ix_2)^m$ equation (10.18) becomes

$$h_{l,m}(x_1, x_2, x_3) = O_l \left[x_3^{l-m}(x_1 \pm ix_2)^m \right]$$

$$= (x_1 \pm ix_2)^m \frac{(l-m)!}{(2l-1)!!} \sum_{j=0}^{[(l-m)/2]} \frac{(-1)^j (2l-2j-1)!!}{(2j)!!(l-m-2j)!} r_{(3)}^{2j} x_3^{l-m-2j}$$

(10.23)

where

$$r_{(3)} \equiv \left(x_1^2 + x_2^2 + x_3^2 \right)^{1/2}$$

(10.24)

The harmonic polynomials shown in equation (10.23) are related to the familiar spherical harmonics by

$$Y_{l,m}(\mathbf{u}) = \frac{\mathcal{N}}{r_{(3)}^l} h_{l,m}(x_1, x_2, x_3)$$

(10.25)

In other words, apart from a constant, the familiar 3-dimensional spherical harmonics can be found by dividing $h_{l,m}(x_1, x_2, x_3)$ by $r_{(3)}^l$. Table 1.2 shows the harmonic polynomials $h_{l,m}(x_1, x_2, x_3)$ for the first few values of $l - m$. The corresponding (unnormalized) spherical harmonics, expressed in terms of spherical polar coordinates, are also shown in the table.

Having found $h_{l,m}(x_1, x_2, x_3)$, we can again make use of equation (10.18). This next step gives us $h_{\lambda,l,m}(x_1, x_2, x_3, x_4)$, a set of harmonic polynomials that are simultaneous eigenfunctions of Λ_4^2, Λ_3^2, and Λ_2^2. For $d = 4$ and $\mu = (l, m)$, equation (10.18) becomes

$$h_{\lambda,l,m}(x_1, x_2, x_3, x_4) = O_\lambda \left[x_4^{\lambda-l} h_{l,m}(x_1, x_2, x_3) \right]$$

$$= h_{lm}(x_1, x_2, x_3) \frac{(\lambda - l)!}{(2l)!!} \sum_{j=0}^{[(\lambda-l)/2]} \frac{(-1)^j (2l-2j)!!}{(2j)!!(\lambda-l-2j)!} r_{(4)}^{2j} x_4^{\lambda-l-2j}$$

(10.26)

where

$$r_{(4)} \equiv \left(x_1^2 + x_2^2 + x_3^2 + x_4^2 \right)^{1/2}$$

(10.27)

Table 10.3 shows $O_\lambda[x_4^{\lambda-l} h_{l,m}(x_1, x_2, x_3)]$ for the first few values of $\lambda - l$. The corresponding (unnormalized) 4-dimensional hyperspherical harmonics are also shown, as functions of the angles χ, θ and ϕ, which are defined by the relationships:

$$u_1 = \frac{x_1}{r_{(4)}} = \sin\chi \sin\theta \cos\phi$$

$$u_2 = \frac{x_2}{r_{(4)}} = \sin\chi \sin\theta \sin\phi$$

$$u_3 = \frac{x_3}{r_{(4)}} = \sin\chi \cos\theta$$

$$u_4 = \frac{x_4}{r_{(4)}} = \cos\chi \tag{10.28}$$

with

$$\sin\chi \equiv \frac{r_{(3)}}{r_{(4)}} \tag{10.29}$$

In Table 10.3, $Y_{l,m}(\theta,\phi)$ are the familiar 3-dimensional spherical harmonics.

10.4 Gegenbauer polynomials

In a 3-dimensional space, the familiar Legendre polynomials are defined in terms of their "generating function", $1/|\mathbf{x}-\mathbf{x}'|$, which is the Green's function of the 3-dimensional Laplacian operator. One first introduces the small parameter ϵ, defined by

$$\epsilon \equiv \frac{r_<}{r_>} \equiv \begin{cases} \dfrac{r}{r'} & \text{if } r \leq r' \\[6pt] \dfrac{r'}{r} & \text{if } r' < r \end{cases} \tag{10.30}$$

Then, expanding the generating function in powers of ϵ, one defines the Legendre polynomial $P_l(\mathbf{u}\cdot\mathbf{u}')$ as the angular function multiplying ϵ^l.

$$\frac{1}{|\mathbf{x}-\mathbf{x}'|} = \frac{1}{r_>(1+\epsilon^2-2\epsilon\mathbf{u}\cdot\mathbf{u}')^{1/2}} = \frac{1}{r_>}\sum_{l=0}^{\infty}\epsilon^l P_l(\mathbf{u}\cdot\mathbf{u}') \tag{10.31}$$

Similarly, in a d-dimensional space, one defines the Gegenbauer polynomials by expanding their generating function in terms of ϵ, where r is now the hyperradius in the d-dimensional space, and the generating function is the Green's function of the generalized Laplacian operator. The Gegenbauer polynomial $C_\lambda^\alpha(\mathbf{u}\cdot\mathbf{u}')$ is defined to be the angular function multiplying ϵ^λ

in the expansion

$$\frac{1}{|\mathbf{x} - \mathbf{x}'|^{d-2}} = \frac{1}{r_>^{d-2}(1 + \epsilon^2 - 2\epsilon \mathbf{u} \cdot \mathbf{u}')^\alpha} = \frac{1}{r_>^{d-2}} \sum_{\lambda=0}^{\infty} \epsilon^\lambda C_\lambda^\alpha(\mathbf{u} \cdot \mathbf{u}') \quad (10.32)$$

where

$$\alpha \equiv \frac{d-2}{2} \quad (10.33)$$

One can see from the way that they are defined that Gegenbauer polynomials are the d-dimensional generalization of Legendre polynomials, and that when $d = 3$ ($\alpha = \frac{1}{2}$), they reduce to Legendre polynomials. From equation (10.32) it is possible to show that

$$C_\lambda^\alpha(\cos \chi) = \frac{1}{(d-4)!!} \sum_{j=0}^{[\lambda/2]} \frac{(-1)^j (d + 2\lambda - 2j - 4)!!}{(2j)!!(\lambda - 2j)!} (\cos \chi)^{\lambda - 2j} \quad (10.34)$$

Just as spherical harmonics can be expressed in terms of Legendre polynomials, hyperspherical harmonics can be expressed in terms of Gegenbauer polynomials. To see this, we can replace α by $\alpha + \mu_1$ in equation (10.34), which is equivalent to replacing d by $d + 2\mu_1$. If we also replace λ by $\lambda - \mu_1$ we obtain

$$C_{\lambda - \mu_1}^{\alpha + \mu_1}(\cos \chi) = \frac{1}{(d + 2\mu_1 - 4)!!}$$
$$\times \sum_{j=0}^{[(\lambda - \mu_1)/2]} \frac{(-1)^j (d + 2\lambda - 2j - 4)!!}{(2j)!!(\lambda - \mu_1 - 2j)!} (\cos \chi)^{\lambda - \mu_1 - 2j}$$

$$(10.35)$$

which can be compared with equations (10.18) and (10.26). It follows from this comparison that we can express the 4-dimensional hyperspherical harmonics in the form

$$Y_{\lambda, l, m}(\mathbf{u}) = \mathcal{N}_{\lambda, l} C_{\lambda - l}^{1 + l}(\cos \chi) \sin^l \chi Y_{l, m}(\theta, \phi) \quad (10.36)$$

where $\mathcal{N}_{\lambda, l}$ is a normalizing constant. The normalizing constant can be evaluated, and it turns out to be

$$\mathcal{N}_{\lambda, l} = (-1)^\lambda i^l (2l)!! \sqrt{\frac{2(\lambda + 1)(\lambda - l)!}{\pi(\lambda + l + 1)!}} \quad (10.37)$$

With this normalization, and with the orthogonality that they are simultaneous eigenfunctions of Λ_4^2, L^2 and L_3, the 4-dimensional hyperspherical harmonics obey the orthonormality relations

$$\int d\Omega \, Y^*_{\lambda',l',m'} Y_{\lambda,l,m} = \delta_{\lambda',\lambda}\delta_{l',l}\delta_{m',m} \tag{10.38}$$

10.5 Hyperspherical expansion of a d-dimensional plane wave

We will next discuss expansions of d-dimensional plane waves in terms of Gegenbauer polynomials, hyperspherical Bessel functions, and hyperspherical harmonics. Such expansions are not only interesting and useful in themselves; they will also help us to discuss conservation of symmetry under Green's function iteration, as well as leading us to an alternative representation of the Green's function. Let us consider a d-dimensional plane wave defined by

$$e^{i\mathbf{p}\cdot\mathbf{x}} \equiv e^{i(p_1 x_1 + \cdots + p_d x_d)} \tag{10.39}$$

Expanding the wave as a Taylor series gives

$$e^{i\mathbf{p}\cdot\mathbf{x}} = \sum_{n=0}^{\infty} \frac{(i\mathbf{p}\cdot\mathbf{x})^n}{n!} = \sum_{n=0}^{\infty} \frac{(ipr)^n(\mathbf{u}_p \cdot \mathbf{u})^n}{n!} \tag{10.40}$$

where

$$\mathbf{u}_p \equiv \frac{1}{p}(p_1, p_2, \ldots, p_d) \tag{10.41}$$

and

$$\mathbf{u} \equiv \frac{1}{r}(x_1, x_2, \ldots, x_d) \tag{10.42}$$

L.K. Hua [Hua, 1963] has shown that

$$(\mathbf{u}_p \cdot \mathbf{u})^n = \frac{n!}{2^n} \sum_{s=0}^{[n/2]} \frac{\alpha + n - 2s}{s!(\alpha)_{n-s+1}} C^\alpha_{n-2s}(\mathbf{u}_p \cdot \mathbf{u}) \tag{10.43}$$

where C^α_λ is a Gegenbauer polynomial,

$$\alpha \equiv \frac{d-2}{2} \tag{10.44}$$

and

$$(\alpha)_j \equiv \alpha(\alpha+1)(\alpha+2)\cdots(\alpha+j-1) \tag{10.45}$$

(see also [Avery, 1989]). Substituting (10.43) into the Taylor series expansion (10.40), we obtain

$$e^{i\mathbf{p}\cdot\mathbf{x}} = \sum_{n=0}^{\infty} \left(\frac{ipr}{2}\right)^n \sum_{s=0}^{[n/2]} \frac{\alpha+n-2s}{s!(\alpha)_{n-s+1}} C_{n-2s}^{\alpha}(\mathbf{u}_p \cdot \mathbf{u}) \qquad (10.46)$$

Finally, isolating the Gegenbauer polynomial corresponding to a particular value of λ and rearranging the summations, we obtain the series

$$e^{i\mathbf{p}\cdot\mathbf{x}} = (d-4)!! \sum_{\lambda=0}^{\infty} i^{\lambda}(d+2\lambda-2) j_{\lambda}^{d}(pr) C_{\lambda}^{\alpha}(\mathbf{u}_p \cdot \mathbf{u}) \qquad (10.47)$$

where the function

$$j_{\lambda}^{d}(pr) \equiv \sum_{t=0}^{\infty} \frac{(-1)^t (pr)^{2t+\lambda}}{(2t)!!(d+2t+2\lambda-2)!!} \qquad (10.48)$$

might be called a "hyperspherical Bessel function", since when $d = 3$ it reduces to a familiar spherical Bessel function. Hyperspherical Bessel functions are related to ordinary Bessel functions by

$$j_{\lambda}^{d}(pr) = \frac{\Gamma(\alpha) 2^{\alpha-1} J_{\alpha+\lambda}(pr)}{(d-4)!!(pr)^{\alpha}} \qquad (10.49)$$

When $d = 3$, (10.47) reduces to the familiar expansion of a plane wave in terms of Legendre polynomials and spherical Bessel functions.

$$e^{i\mathbf{p}\cdot\mathbf{x}} = \sum_{l=0}^{\infty} i^l (2l+1) j_l(pr) P_l(\mathbf{u}_p \cdot \mathbf{u}) \qquad (10.50)$$

In fact, almost all of the beautiful theorems that can be established for spherical harmonics and Legendre polynomials have d-dimensional counterparts.

Hyperspherical harmonics obey the sum rule ([Avery, 1989])

$$C_{\lambda}^{\alpha}(\mathbf{u}_p \cdot \mathbf{u}) = \frac{(d-2) I(0)}{(d+2\lambda-2)} \sum_{\mu} Y_{\lambda,\mu}^{*}(\mathbf{u}_p) Y_{\lambda,\mu}(\mathbf{u}) \qquad (10.51)$$

where $I(0)$ is the total generalized solid angle in a d-dimensional space:

$$I(0) \equiv \int d\Omega = \frac{2\pi^{d/2}}{\Gamma(d/2)} \qquad (10.52)$$

When $d = 3$, equation (10.51) reduces to the familiar sum rule relating spherical harmonics to Legendre polynomials:

$$\frac{4\pi}{2l+1} \sum_m Y_{lm}^*(\mathbf{u}')Y_{lm}(\mathbf{u}) = P_l(\mathbf{u} \cdot \mathbf{u}') \tag{10.53}$$

In equation (10.51) the hyperspherical harmonics are assumed to be mutually orthogonal and normalized, and the sum $\sum_\mu$ is taken over all of the hyperspherical harmonics corresponding to a particular value of the principal quantum number λ. Equation (10.51) implies that if $F(\mathbf{u})$ is any function of the angles in a d-dimensional space, then

$$O_\lambda[F(\mathbf{u})] = \frac{\Gamma(d/2)(2\lambda+d-2)}{2\pi^{\frac{d}{2}}(d-2)} \int d\Omega' \, C_\lambda^\alpha(\mathbf{u} \cdot \mathbf{u}')F(\mathbf{u}') \tag{10.54}$$

projects out the component of $F(\mathbf{u})$ that is an eigenfunction of Λ_d^2. From equation (10.51) it also follows that the Gegenbauer polynomials with argument $\mathbf{u}_p \cdot \mathbf{u}$ are eigenfunctions of the generalized angular momentum operator Λ^2 in the d-dimensional space:

$$\Lambda^2 C_\lambda^\alpha(\mathbf{u}_p \cdot \mathbf{u}) = \lambda(\lambda+d-2)C_\lambda^\alpha(\mathbf{u}_p \cdot \mathbf{u}) \tag{10.55}$$

since the hyperspherical harmonics $Y_{\lambda,\mu}(\mathbf{u})$ are eigenfunctions of this operator, and all the harmonics that enter the sum over μ correspond to the same value of λ. Thus equation (10.47) can be interpreted as the expansion of a d-dimensional plane wave in terms of eigenfunctions of generalized angular momentum. If O_λ is a projection operator corresponding to a particular value of λ it follows that

$$O_\lambda\left[e^{i\mathbf{p}\cdot\mathbf{x}}\right] = (d-4)!!i^\lambda(d+2\lambda-2)j_\lambda^d(pr)C_\lambda^\alpha(\mathbf{u}_p \cdot \mathbf{u}) \tag{10.56}$$

We can also use equation (10.51) to rewrite (10.47) in the form

$$e^{i\mathbf{p}\cdot\mathbf{x}} = (d-2)!!I(0) \sum_{\lambda=0}^\infty i^\lambda j_\lambda^d(pr) \sum_\mu Y_{\lambda,\mu}^*(\mathbf{u}_p)Y_{\lambda,\mu}(\mathbf{u}) \tag{10.57}$$

The equation analogous to (10.56) becomes

$$O_\lambda\left[e^{i\mathbf{p}\cdot\mathbf{x}}\right] = (d-2)!!I(0)i^\lambda j_\lambda^d(pr) \sum_\mu Y_{\lambda,\mu}^*(\mathbf{u}_p)Y_{\lambda,\mu}(\mathbf{u}) \tag{10.58}$$

and in particular

$$O_0\left[e^{i\mathbf{p}\cdot\mathbf{x}}\right] = \frac{1}{I(0)} \int d\Omega \, e^{i\mathbf{p}\cdot\mathbf{x}} = (d-2)!!j_0^d(pr) \tag{10.59}$$

10.6 Alternative hyperspherical harmonics; The method of trees

Professor V. Aquilanti and his coworkers at the University of Perugia, together with a number of Russian authors, have studied alternative ways of constructing orthonormal sets of hyperspherical harmonics. The method developed by these authors is called the *method of trees* [Aquilanti et al., 1996], [Coletti, 1998]. To illustrate it, we can think of a d-dimensional space, divided into two subspaces, whose dimensions are respectively d_1 and d_2, with $d_1 + d_2 = d$. Let $\mathbf{x}_1$ represent the Cartesian coordinate vector of points in the first subspace, while $\mathbf{x}_2$ is the corresponding vector for points in the second subspace. The Cartesian coordinate vector $\mathbf{x}$ for the entire d-dimensional space is thus

$$\mathbf{x} = (\mathbf{x}_1, \mathbf{x}_2) \tag{10.60}$$

We also define r_1, r_2 and r by the relationships

$$r_1^2 = \mathbf{x}_1 \cdot \mathbf{x}_1$$
$$r_2^2 = \mathbf{x}_2 \cdot \mathbf{x}_2$$
$$r^2 = \mathbf{x} \cdot \mathbf{x} \tag{10.61}$$

Now let $h_{l_1}(\mathbf{x}_1)$ be a harmonic polynomial of degree l_1 in the first subspace, while $h_{l_2}(\mathbf{x}_2)$ is a harmonic polynomial of degree l_2 in the second subspace. Then

$$\Delta[h_{l_1}(\mathbf{x}_1)h_{l_2}(\mathbf{x}_2)] = h_{l_2}(\mathbf{x}_2)\Delta_1 h_{l_1}(\mathbf{x}_1) + h_{l_1}(\mathbf{x}_1)\Delta_2 h_{l_2}(\mathbf{x}_2) = 0 \tag{10.62}$$

where

$$\Delta = \Delta_1 + \Delta_2$$
$$\Delta_1 = \sum_{j=1}^{d_1} \frac{\partial^2}{\partial x_j^2}$$
$$\Delta_2 = \sum_{j=d_1+1}^{d} \frac{\partial^2}{\partial x_j^2} \tag{10.63}$$

If β and β' are even integers, and $\lambda = \beta + \beta' + l_1 + l_2$, then

$$f_{\lambda, l_1, l_2}(\mathbf{x}) = r_1^\beta r_2^{\beta'} h_{l_1}(\mathbf{x}_1) h_{l_2}(\mathbf{x}_2) \tag{10.64}$$

will be a homogeneous polynomial of degree λ in the coordinates $x_1, x_2, \ldots, x_d$. For example,

$$f_{\lambda, l_1, l_2}(\mathbf{x}) = r_1^2 h_{l_1}(\mathbf{x}_1) h_{l_2}(\mathbf{x}_2) \tag{10.65}$$

is a homogeneous polynomial of degree $\lambda = l_1 + l_2 + 2$ in the d-dimensional space. Applying the generalized Laplacian operator Δ to the homogeneous polynomial shown in equation (10.65), we obtain:

$$\Delta[r_1^2 h_{l_1}(\mathbf{x}_1) h_{l_2}(\mathbf{x}_2)] = h_{l_2}(\mathbf{x}_2) \Delta_1[r_1^2 h_{l_1}(\mathbf{x}_1)]$$

$$= h_{l_2}(\mathbf{x}_2) \left[2 d_1 h_{l_1}(\mathbf{x}_1) + 2 \sum_{j=1}^{d_1} x_j \frac{\partial h_{l_1}(\mathbf{x}_1)}{\partial x_j} \right]$$

$$= 2(d_1 + 2l_1) h_{l_1}(\mathbf{x}_1) h_{l_2}(\mathbf{x}_2) \qquad (10.66)$$

A second application of the generalized Laplacian operator gives zero:

$$\Delta^2[r_1^2 h_{l_1}(\mathbf{x}_1) h_{l_2}(\mathbf{x}_2)] = 0 \qquad (10.67)$$

Then, using equations (10.66), (10.67) and (9.32), we can construct the harmonic polynomial of highest degree in the canonical decomposition of $r_1^2 h_{l_1}(\mathbf{x}_1) h_{l_2}(\mathbf{x}_2)$:

$$h_{\lambda, l_1, l_2}(\mathbf{x}) = \left[r_1^2 - \frac{(d_1 + 2l_1)}{(d + 2\lambda - 4)} r^2 \right] h_{l_1}(\mathbf{x}_1) h_{l_2}(\mathbf{x}_2) \qquad (10.68)$$

This can be rewritten in a more symmetrical form:

$$h_{\lambda, l_1, l_2}(\mathbf{x}) = \left[\frac{(d_2 + 2l_2) r_1^2 - (d_1 + 2l_1) r_2^2}{(d + 2\lambda - 4)} \right] h_{l_1}(\mathbf{x}_1) h_{l_2}(\mathbf{x}_2) \qquad (10.69)$$

Equation (10.69), and analogous equations with higher values of β and β', can be used to construct the alternative set of 4-dimensional hyperspherical harmonics shown in Table 5. To do so, we let $d_1 = d_2 = 2$, $d = 4$, with

$$r_1^2 = x_1^2 + x_2^2, \qquad r_2^2 = x_3^2 + x_4^2 \qquad (10.70)$$

and

$$h_{l_1}(\mathbf{x}_1) = (x_1 \pm i x_2)^{l_1} \qquad h_{l_2}(\mathbf{x}_2) = (x_3 \pm i x_4)^{l_2} \qquad (10.71)$$

The reader may enjoy constructing the first few 4-dimensional hyperspherical harmonics of this type and comparing the results with Table 10.5. The alternative hyperspherical harmonics shown in Table 10.5 correspond, through Fock's projection, to the direct-space hydrogenlike orbitals that are obtained using parabolic coordinates:

$$\xi \equiv r + z$$
$$\eta \equiv r - z$$

$$\varphi \equiv \tan^{-1}\left(\frac{y}{x}\right) \tag{10.72}$$

The Schrödinger equation for hydrogenlike atoms is separable in parabolic coordinates, and the direct-space solutions have the form:

$$w_{n_1,n_2,m}(\mathbf{x}) = \sqrt{\frac{n_1! n_2! k^{2|m|+3}}{\pi n[(n_1+|m|)!(n_2+|m|)!]^3}}$$
$$\times e^{-k(\xi+\eta)/2}(\xi\eta)^{|m|/2} e^{im\varphi} L_{n_1+|m|}^{|m|}(k\xi) L_{n_2+|m|}^{|m|}(k\eta) \tag{10.73}$$

where $n = n_1 + n_2 + m + 1$ and m play their usual roles as the principal quantum number and the magnetic quantum number respectively, and where the functions $L_p^q(\rho)$ are the associated Laguerre polynomials defined by

$$L_p^q(\rho) = \frac{d^q}{d\rho^q}\left[e^{\rho}\frac{d^p}{d\rho^p}\left(e^{-\rho}\rho^p\right)\right] \tag{10.74}$$

The Fourier transform of $w_{n_1,n_2,m}(\mathbf{x})$ is related to the alternative hyperspherical harmonics of Table 10.5 by

$$w_{n_1,n_2,m}^t(\mathbf{p}) = M(p)Y_{n_1,n_2,m}(\mathbf{u}) \tag{10.75}$$

where $M(p)$ is defined by equation (8.21) while $\mathbf{u}$ is defined by (8.19).

The method of trees takes its name from the tree-like graphical representations of the chains of subgroups used to organize the minor indices of hyperspherical harmonics. The indices of the standard set of 4-dimensional hyperspherical harmonics shown in Table 10.4 are organized according to the chain of subgroups

$$SO(4) \supset SO(3) \supset SO(2) \tag{10.76}$$

while the indices of the alternative harmonics of Table 10.5 are organized according to the chain

$$SO(4) \supset SO(2) \times SO(2) \tag{10.77}$$

These two chains are symbolized by the tree-like graphs shown in Figure 10.1. On the left of the figure, the standard tree is shown. It symbolizes the procedure that we followed in equations (10.19)-(10.27), where we first found harmonic polynomials in the subspace spanned by x_1 and x_2, then multiplied these by x_3 raised to some power and projected out the harmonic part of the resulting homogeneous polynomial, and finally repeated the procedure for x_4. By contrast, the alternative tree at the right side of

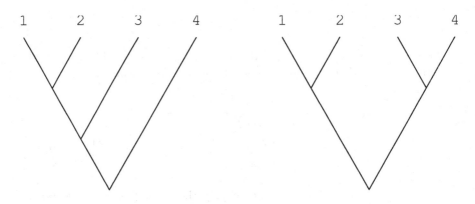

Fig. 10.1 The standard tree (left) and an alternative tree (right)

the figure symbolizes the procedure where one first constructs a harmonic polynomial in the subspace spanned by x_1 and x_2, and another harmonic polynomial in the subspace spanned by x_3 and x_4, the final step being to use these two building blocks to construct harmonic polynomials in the entire space.

The authors mentioned above, who developed the method of trees, have found ways of treating a general fork in a d-dimensional space. They have shown that if

$$f_{\lambda,l_1,l_2}(\mathbf{x}) = x_j^{\lambda-l_1-l_2} h_{l_1}(\mathbf{x}_1) h_{l_2}(\mathbf{x}_2) \tag{10.78}$$

is a homogeneous polynomial of degree λ in a d-dimensional space, and if $h_{l_1}(\mathbf{x}_1)$ and $h_{l_2}(\mathbf{x}_2)$ are harmonic two subspaces of which together form the entire space, then the harmonic polynomial of highest degree in the canonical decomposition of $f_{\lambda,l_1,l_2}(\mathbf{x})$ is given by

$$h_{\lambda,l_1,l_2}(\mathbf{x}) \sim r^{\lambda-l_1-l_2} P_{(\lambda-l_1-l_2)/2}^{(l_2+\alpha_2,l_1+\alpha_1)} \left(\frac{r_1^2 - r_2^2}{r^2} \right) h_{l_1}(\mathbf{x}_1) h_{l_2}(\mathbf{x}_2) \tag{10.79}$$

where $P_n^{(a,b)}$ is a Jacobi polynomial and where

$$\alpha_j \equiv \frac{d_j - 2}{2} \qquad j = 1, 2 \tag{10.80}$$

d_1 and d_2 being the dimensions of the subspaces (see [Coletti, 1998]).

The theory of harmonic polynomials and hyperspherical harmonics can be seen to offer a valuable supplement to group-theoretical methods for dealing with angular momentum and hyperangular momentum, as well as with angular and hyperangular integrations. We can also see that the theory is closely related to the momentum-space representation of Sturmians.

These momentum-space relationships are utilized in Chapter 11 to discuss the extension of Sturmian methods to many-center problems. In Appendix A, we shall use harmonic projection to evaluate the angular integrals needed for relativistic calculations, while in Appendix E, we shall use the theory of hyperspherical harmonics to discuss conservation of symmetry under iteration of the many-electron Schrödinger equation.

Table 10.1: Equation (10.18) for various values of $\lambda - \mu_1$.

$\lambda - \mu_1$	$\mathcal{O}_\lambda \left[x_d^{\lambda-\mu_1} h_\mu(x_1, \ldots, x_{d-1}) \right]$
0	$h_\mu(x_1, \ldots, x_{d-1})$
1	$x_d h_\mu(x_1, \ldots, x_{d-1})$
2	$\left(x_d^2 - \dfrac{r^2}{d+2\lambda-4} \right) h_\mu(x_1, \ldots, x_{d-1})$
3	$\left(x_d^3 - \dfrac{3 x_d r^2}{d+2\lambda-4} \right) h_\mu(x_1, \ldots, x_{d-1})$

Table 10.2: Equation (10.18) in the special case where $d = 3$, $\lambda = l$, $\mu_1 = m$, and $h_\mu(x_1, \ldots, x_{d-1}) = (x_1 + ix_2)^m$. The familiar 3-dimensional spherical harmonics (unnormalized) are shown in the right column.

$l - m$	$O_l\left[x_3^{l-m}(x_1 + ix_2)^m\right]$	$Y_{l,m}(\theta, \phi)$
0	$(x_1 + ix_2)^m$	$\sin^m \theta e^{im\phi}$
1	$x_3(x_1 + ix_2)^m$	$\cos\theta \sin^m \theta e^{im\phi}$
2	$\left(x_3^2 - \dfrac{r_{(3)}^2}{2l+1}\right)(x_1 + ix_2)^m$	$\left(\cos^2\theta - \dfrac{1}{2l+1}\right)\sin^m \theta e^{im\phi}$
3	$\left(x_3^3 - \dfrac{3x_3 r_{(3)}^2}{2l+1}\right)(x_1 + ix_2)^m$	$\left(\cos^3\theta - \dfrac{3\cos\theta}{2l+1}\right)\sin^m \theta e^{im\phi}$

Table 10.3: Equation (10.18) in the special case where $d = 4$, $\mu_1 = l$ and $h_\mu(x_1, \ldots, x_{d-1}) = h_{l,m}(x_1, x_2, x_3)$. The (unnormalized) 4-dimensional hyperspherical harmonics are shown in the right column.

$\lambda - l$	$O_\lambda \left[x_4^{\lambda-l} h_{l,m}(x_1, x_2, x_3) \right]$	$Y_{\lambda,l,m}(\chi, \theta, \phi)$
0	$h_{l,m}(x_1, x_2, x_3)$	$\sin^l \chi Y_{l,m}(\theta, \phi)$
1	$x_4 h_{l,m}(x_1, x_2, x_3)$	$\cos \chi \sin^l \chi Y_{l,m}(\theta, \phi)$
2	$\left(x_4^2 - \dfrac{r_{(4)}^2}{2\lambda} \right) h_{l,m}(x_1, x_2, x_3)$	$\left(\cos^2 \chi - \dfrac{1}{2\lambda} \right) \sin^l \chi Y_{l,m}(\theta, \phi)$
3	$\left(x_4^3 - \dfrac{3 x_4 r_{(4)}^2}{2\lambda} \right) h_{l,m}(x_1, x_2, x_3)$	$\left(\cos^3 \chi - \dfrac{3 \cos \chi}{2\lambda} \right) \sin^l \chi Y_{l,m}(\theta, \phi)$

Table 10.4: 4-dimensional hyperspherical harmonics

λ	l	m	$\sqrt{2\pi}\ Y_{\lambda,l,m}(\mathbf{u})$
0	0	0	1
1	1	1	$i\sqrt{2}(u_1+iu_2)$
1	1	0	$-i2u_3$
1	1	-1	$-i\sqrt{2}(u_1-iu_2)$
1	0	0	$-2u_4$

λ	l	m	$\sqrt{2\pi}\ Y_{\lambda,l,m}(\mathbf{u})$
2	2	2	$-\sqrt{3}(u_1+iu_2)^2$
2	2	1	$2\sqrt{3}u_3(u_1+iu_2)$
2	2	0	$-\sqrt{2}(2u_3^2-u_1^2-u_2^2)$
2	2	-1	$-2\sqrt{3}u_3(u_1-iu_2)$
2	2	-2	$-\sqrt{3}(u_1-iu_2)^2$
2	1	1	$-i2\sqrt{3}\ u_4(u_1+iu_2)$
2	1	0	$2i\sqrt{6}\ u_4 u_3$
2	1	-1	$2i\sqrt{3}\ u_4(u_1-iu_2)$
2	0	0	$4u_4^2-1$

Table 4 continued.

λ	l	m	$\sqrt{2}\pi\, Y_{\lambda,l,m}(\mathbf{u})$
3	3	3	$-2i(u_1 + iu_2)^3$
3	3	2	$i\sqrt{24}u_3(u_1 + iu_2)^2$
3	3	1	$2i\sqrt{\dfrac{3}{5}}(u_1 + iu_2)(u_1^2 + u_2^2 - 4u_3^2)$
3	3	0	$-\dfrac{4i}{\sqrt{5}}u_3(3u_1^2 + 3u_2^2 - 2u_3^2)$
3	2	2	$\sqrt{24}u_4(u_1 + iu_2)^2$
3	2	1	$-\sqrt{96}(u_1 + iu_2)u_3 u_4$
3	2	0	$-4u_4(u_1^2 + u_2^2 - 2u_3^2)$
3	1	1	$\sqrt{\dfrac{8}{5}}(u_1 + iu_2)(6u_4^2 - 1)$
3	1	0	$i\dfrac{4}{\sqrt{5}}u_3(6u_4^2 - 1)$
3	0	0	$4u_4(1 - 2u_4^2)$

Table 10.5: Hyperspherical harmonics corresponding to solutions for the hydrogen atom in parabolic coordinates, with $\mathbf{t} \equiv k\mathbf{x}$ and $t \equiv kr$.

n_1	n_2	m	$\sqrt{2}\pi\, \mathcal{Y}_{n_1,n_2,m}(\mathbf{u})$	$\left(\pi/k^3\right)^{1/2} w_{n_1,n_2,m}(\mathbf{x})$
0	0	0	1	e^{-t}
0	0	1	$-i\sqrt{2}(u_1 + iu_2)$	$\dfrac{1}{\sqrt{2}}\, e^{-t}(t_1 + it_2)$
0	0	-1	$-i\sqrt{2}(u_1 - iu_2)$	$\dfrac{1}{\sqrt{2}}\, e^{-t}(t_1 - it_2)$
1	0	0	$\sqrt{2}(u_4 + iu_3)$	$\dfrac{1}{\sqrt{2}}\, e^{-t}(1 - t - t_3)$
0	1	0	$\sqrt{2}(u_4 - iu_3)$	$\dfrac{1}{\sqrt{2}}\, e^{-t}(1 - t + t_3)$

Table 5 continued.

n_1	n_2	m	$\sqrt{2}\pi\, \mathcal{Y}_{n_1,n_2,m}(\mathbf{u})$
0	0	2	$-\sqrt{3}(u_1+iu_2)^2$
0	0	-2	$-\sqrt{3}(u_1-iu_2)^2$
2	0	0	$\sqrt{3}(u_4+iu_3)^2$
0	2	0	$\sqrt{3}(u_4-iu_3)^2$
1	1	0	$\sqrt{3}(u_1^2+u_2^2-u_3^2-u_4^2)$
1	0	1	$\sqrt{6}(u_4+iu_3)(u_1+iu_2)$
1	0	-1	$\sqrt{6}(u_4+iu_3)(u_1-iu_2)$
0	1	1	$\sqrt{6}(u_4-iu_3)(u_1+iu_2)$
0	1	-1	$\sqrt{6}(u_4-iu_3)(u_1-iu_2)$

Table 10.6: Gegenbauer polynomials. $(\alpha)_j \equiv \alpha(\alpha+1)(\alpha+2)\cdots(\alpha+j-1)$.

λ	$C_\lambda^\alpha(\zeta)$	$C_\lambda^1(\zeta)$
0	1	1
1	$2\alpha\zeta$	2ζ
2	$2(\alpha)_2\zeta^2 - \alpha$	$4\zeta^2 - 1$
3	$\dfrac{1}{3}\left[4(\alpha)_3\zeta^3 - 6(\alpha)_2\zeta\right]$	$8\zeta^3 - 4\zeta$
4	$\dfrac{1}{6}\left[4(\alpha)_4\zeta^4 - 12(\alpha)_3\zeta^2 + 3(\alpha)_2\right]$	$16\zeta^4 - 12\zeta^2 + 1$
5	$\dfrac{1}{15}\left[4(\alpha)_5\zeta^5 - 20(\alpha)_4\zeta^3 + 15(\alpha)_3\zeta\right]$	$32\zeta^5 - 32\zeta^3 + 6\zeta$

Chapter 11

THE MANY-CENTER PROBLEM

11.1 The many-center one-electron problem

Although this book deals with the spectra of atoms and ions, it may be interesting for the reader to see how Sturmian methods can be applied to many-center problems. Thus we will end the book with a few remarks about the application of Sturmian methods to molecules and molecular ions.

Let us consider an electron moving in the attractive Coulomb potential of a number of nuclei. The wave function of the electron, $\varphi_j(\mathbf{x})$, will be a solution of

$$\left[-\frac{1}{2}\nabla^2 + v(\mathbf{x}) - \epsilon_j\right]\varphi_j(\mathbf{x}) = 0 \qquad (11.1)$$

where

$$v(\mathbf{x}) = -\sum_a \frac{Z_a}{|\mathbf{x} - \mathbf{X}_a|} \qquad (11.2)$$

If we define the set of indices τ by

$$\tau \equiv (n, l, m, a) \qquad (11.3)$$

we can express the solutions to (11.1) in the form

$$\varphi_j(\mathbf{x}) = \sum_\tau \chi_\tau(\mathbf{x}) C_{\tau,j} \qquad (11.4)$$

where the set of functions

$$\chi_\tau(\mathbf{x}) \equiv \chi_{nlm}(\mathbf{x} - \mathbf{X}_a) \qquad (11.5)$$

are one-electron Coulomb Sturmians located on the various centers $\mathbf{X}_a$. The Coulomb Sturmians are spherical harmonics multiplied by radial functions of the type shown in Table 1.1. The exponential factor e^{-kr} common to all the radial functions is related to the energies ϵ_j by

$$\epsilon_j \equiv -\frac{1}{2}k^2 \qquad (11.6)$$

With the help of (11.4) and (11.6), equation (11.1) can be rewritten in the form

$$\sum_\tau \left[-\frac{1}{2}\nabla^2 + \frac{1}{2}k^2 + v(\mathbf{x})\right]\chi_\tau(\mathbf{x})C_{\tau,j} = 0 \qquad (11.7)$$

11.2 Shibuya-Wulfman integrals

The basis functions shown in equation (11.5) can be expressed in terms of their Fourier transforms:

$$\chi_\tau(\mathbf{x}) = \frac{1}{(2\pi)^{3/2}}\int d^3p\, e^{i\mathbf{p}\cdot\mathbf{x}}\chi_\tau^t(\mathbf{p}) \qquad (11.8)$$

Since the Laplacian operator acting on a plane wave brings down a factor of $-p^2$, it follows from (11.8) that

$$\left(\frac{-\nabla^2 + k^2}{2k^2}\right)\chi_\tau(\mathbf{x}) = \frac{1}{(2\pi)^{3/2}}\int d^3p\, e^{i\mathbf{p}\cdot\mathbf{x}}\left(\frac{k^2 + p^2}{2k^2}\right)\chi_\tau^t(\mathbf{p}) \qquad (11.9)$$

If we multiply both sides of (11.9) from the left by a conjugate basis function, integrate over d^3x, we obtain:

$$\int d^3x\, \chi_{\tau'}^*(\mathbf{x})\left(\frac{-\nabla^2 + k^2}{2k^2}\right)\chi_\tau(\mathbf{x})$$

$$= \frac{1}{(2\pi)^{3/2}}\int d^3x\, \chi_{\tau'}^*(\mathbf{x})\int d^3p\, e^{i\mathbf{p}\cdot\mathbf{x}}\left(\frac{k^2+p^2}{2k^2}\right)\chi_\tau^t(\mathbf{p})$$

$$= \int d^3p\left(\frac{1}{(2\pi)^{3/2}}\int d^3x\, e^{-i\mathbf{p}\cdot\mathbf{x}}\chi_{\tau'}^t(\mathbf{x})\right)^*\left(\frac{k^2+p^2}{2k^2}\right)\chi_\tau^t(\mathbf{p})$$

$$= \int d^3p\, \chi_{\tau'}^{t*}(\mathbf{p})\left(\frac{k^2+p^2}{2k^2}\right)\chi_\tau^t(\mathbf{p}) \equiv \mathfrak{S}_{\tau',\tau} \qquad (11.10)$$

Integrals of the type shown in equation (11.10) were first studied by T. Shibuya and C.E. Wulfman [Shibuya and Wulfman, 1965], who generalized V. Fock's momentum-space methods to many-center problems. Comparing equation (11.10) with (2.23) and (2.24), we can see that when the two

Coulomb Sturmians are located on the same center, the Shibuya-Wulfman integral $\mathfrak{S}_{\tau',\tau}$ reduces to a delta-function in the remaining indices of τ:

$$\mathfrak{S}_{\tau',\tau} = \delta_{n',n}\delta_{l',l}\delta_{m',m} \qquad \text{if } a' = a \qquad (11.11)$$

Making use of the Fock projection (Appendix D), we can express the Fourier transforms of our basis functions in terms of 4-dimensional hyperspherical harmonics:

$$\chi_\tau^t(\mathbf{p}) = e^{-i\mathbf{p}\cdot\mathbf{X}_a} \chi_{nlm}^t(\mathbf{p}) = e^{-i\mathbf{p}\cdot\mathbf{X}_a} \frac{4k^{5/2}}{(k^2+p^2)^2} Y_{n-1,l,m}(\mathbf{u}) \qquad (11.12)$$

Since, with the Fock projection, the momentum-space volume element d^3p is related to $d\Omega$ by

$$d^3p = \left(\frac{k^2+p^2}{2k}\right) d\Omega \qquad (11.13)$$

we can also write the Shibuya-Wulfman integrals in the form:

$$\mathfrak{S}_{\tau',\tau} = \int d\Omega \, e^{i\mathbf{p}\cdot(\mathbf{X}_{a'}-\mathbf{X}_a)} Y^*_{n'-1,l',m'}(\mathbf{u}) Y_{n-1,l,m}(\mathbf{u}) \qquad (11.14)$$

(from which it can be seen that (11.11) is satisfied) and this is the form in which the integrals were originally studied by Shibuya and Wulfman. To evaluate the integrals, these two authors made an expansion of the plane wave $e^{i\mathbf{p}\cdot(\mathbf{X}_{a'}-\mathbf{X}_a)}$ in terms of one-electron Coulomb Sturmians and their Fourier transforms:

$$e^{i\mathbf{p}\cdot\mathbf{R}} = (2\pi)^{3/2} \left(\frac{p^2+k^2}{2k^2}\right) \sum_{nlm} \chi_{nlm}^{t*}(\mathbf{p})\chi_{nlm}(\mathbf{R}) \qquad (11.15)$$

which is of the form shown in equation (2.28). Remembering from Appendix D that

$$\chi_{nlm}^t(\mathbf{p}) = \frac{4k^{5/2}}{(k^2+p^2)^2} Y_{n-1,l,m}(\mathbf{u}) \qquad (11.16)$$

and that

$$u_4 = \frac{k^2-p^2}{k^2+p^2} \qquad (11.17)$$

we can rewrite the plane wave expansion of equation (11.15) as

$$e^{i\mathbf{p}\cdot\mathbf{R}} = \left(\frac{2\pi}{k}\right)^{3/2} (1+u_4) \sum_{nlm} \chi_{nlm}(\mathbf{R}) Y^*_{n-1,l,m}(\mathbf{u}) \qquad (11.18)$$

The Shibuya-Wulfman integrals can then be expressed in the form

$$\mathfrak{S}_{\tau',\tau} = \left(\frac{2\pi}{k}\right)^{3/2} \sum_{n''l''m''} \chi_{n''l''m''}(\mathbf{X}_{a'} - \mathbf{X}_a)$$

$$\times \int d\Omega \; (1+u_4) Y^*_{n''-1,l'',m''}(\mathbf{u}) Y^*_{n'-1,l',m'}(\mathbf{u}) Y_{n-1,l,m}(\mathbf{u})$$
(11.19)

which allows the integrals to be evaluated using the hyperangular integration methods discussed in Chapter 9. Alternatively the Shibuya-Wulfman integrals can be expressed in terms of Clebsch-Gordan coefficients and 9j symbols, as has been shown by V. Aquilanti and A. Caligiana [Aquilanti and Caligiana, 2002]-[Aquilanti and Caligiana, 2003], [Caligiana, 2003]. Caligiana has also written an efficient FORTRAN program for the generation of Shibuya-Wulfman integrals, and an independent program for generating these integrals has been written by C. Weatherford at the University of Florida.

11.3 Shibuya-Wulfman integrals and translations

As V. Aquilanti has shown, the Shibuya-Wulfman integrals can be related to the effect of a translation on Coulomb Sturmians. To see this we can begin by expressing the basis function χ_τ in terms of its Fourier transform:

$$\chi_{nlm}(\mathbf{x}-\mathbf{X}_a) = \frac{1}{(2\pi)^{3/2}} \int d^3p \; e^{i\mathbf{p}\cdot(\mathbf{x}-\mathbf{X}_a)} \chi^t_{nlm}(\mathbf{p})$$
(11.20)

We then replace $\mathbf{R}$ by $\mathbf{x} - \mathbf{X}_{a'}$ in the Sturmian expansion of a plane wave shown in equation (11.15):

$$e^{i\mathbf{p}\cdot(\mathbf{x}-\mathbf{X}_{a'})} = (2\pi)^{3/2} \left(\frac{k^2+p^2}{2k^2}\right) \sum_{n'l'm'} \chi^{t*}_{n'l'm'}(\mathbf{p}) \chi_{n'l'm'}(\mathbf{x}-\mathbf{X}_{a'})$$ (11.21)

This can of course be rewritten in the form:

$$e^{i\mathbf{p}\cdot\mathbf{x}} = (2\pi)^{3/2} \left(\frac{k^2+p^2}{2k^2}\right) e^{i\mathbf{p}\cdot\mathbf{X}_{a'}} \sum_{n'l'm'} \chi^{t*}_{n'l'm'}(\mathbf{p}) \chi_{n'l'm'}(\mathbf{x}-\mathbf{X}_{a'})$$ (11.22)

Replacing $e^{i\mathbf{p}\cdot\mathbf{x}}$ in (11.20) by this expansion, we obtain

$$\chi_{n,l,m}(\mathbf{x}-\mathbf{X}_a) = \sum_{n'l'm'} \chi_{n'l'm'}(\mathbf{x}-\mathbf{X}_{a'})$$

$$\times \int d^3p \left(\frac{k^2 + p^2}{2k^2} \right) e^{i\mathbf{p}\cdot(\mathbf{X}_{a'} - \mathbf{X}_a)} \chi_{n'l'm'}^{t*}(\mathbf{p}) \chi_{nlm}^{t}(\mathbf{p}) \quad (11.23)$$

If we compare (11.23) with (11.10), we can see that

$$\chi_{nlm}(\mathbf{x} - \mathbf{X}_a) = \sum_{n'l'm'} \chi_{n'l'm'}(\mathbf{x} - \mathbf{X}_{a'}) \mathfrak{S}_{\tau',\tau} \quad (11.24)$$

or, more simply,

$$\chi_\tau(\mathbf{x}) = \sum_{n'l'm'} \chi_{\tau'}(\mathbf{x}) \mathfrak{S}_{\tau',\tau} \quad (11.25)$$

In other words, if a Coulomb Sturmian located on one center is expanded in terms of Coulomb Sturmians located on another center, the expansion coefficients are Shibuya-Wulfman integrals[1]. Since two translations performed in succession can be expressed as a single translation, we have

$$\sum_{n'l'm'} \mathfrak{S}_{\tau'',\tau'} \mathfrak{S}_{\tau',\tau} = \mathfrak{S}_{\tau'',\tau} \quad (11.26)$$

Another property of $\mathfrak{S}_{\tau',\tau}$ follows from equation (11.10):

$$\mathfrak{S}_{\tau',\tau}^* = \mathfrak{S}_{\tau,\tau'} \quad (11.27)$$

11.4 Matrix elements of the nuclear attraction potential

From equation (1.15) it follows that

$$\left[-\frac{1}{2}\nabla^2 + \frac{1}{2}k^2 - \frac{nk}{|\mathbf{x} - \mathbf{X}_a|} \right] \chi_{nlm}(\mathbf{x} - \mathbf{X}_a) = 0 \quad (11.28)$$

and combining (11.28) with (11.10) yields

$$\mathfrak{S}_{\tau',\tau} = \frac{n}{k} \int d^3x \, \chi_{\tau'}^*(\mathbf{x}) \frac{1}{|\mathbf{x} - \mathbf{X}_a|} \chi_\tau(\mathbf{x})$$
$$\equiv \frac{n}{k} \int d^3x \, \chi_{n'l'm'}^*(\mathbf{x} - \mathbf{X}_{a'}) \frac{1}{|\mathbf{x} - \mathbf{X}_a|} \chi_{nlm}(\mathbf{x} - \mathbf{X}_a) \quad (11.29)$$

Finally, from (11.28) and (11.24) we obtain the three-center integral

$$\sum_a \int d^3x \, \chi_{n'l'm'}^*(\mathbf{x} - \mathbf{X}_{a'}) \frac{Z_a}{|\mathbf{x} - \mathbf{X}_a|} \chi_{n''l''m''}(\mathbf{x} - \mathbf{X}_{a''})$$

---

[1] It should be noted, however, that this expansion is very slowly convergent when $k|\mathbf{X}_{a'} - \mathbf{X}_a|$ is large.

$$= \sum_{nlma} \mathfrak{S}_{\tau,\tau''} \int d^3x\, \chi^*_{n'l'm'}(\mathbf{x}-\mathbf{X}_{a'}) \frac{Z_a}{|\mathbf{x}-\mathbf{X}_a|} \chi_{nlm}(\mathbf{x}-\mathbf{X}_a)$$

$$= \sum_\tau \mathfrak{S}_{\tau',\tau} \frac{Z_a k}{n} \mathfrak{S}_{\tau,\tau''} \qquad (11.30)$$

This result can be used to evaluate matrix elements of the many-center potential $v(\mathbf{x})$ of equation (11.2):

$$\mathfrak{W}_{\tau',\tau''} \equiv -\frac{1}{k} \int d^3x\, \chi^*_{\tau'}(\mathbf{x}) v(\mathbf{x}) \chi_{\tau''}(\mathbf{x}) = \sum_\tau \mathfrak{S}_{\tau',\tau} \frac{Z_a}{n} \mathfrak{S}_{\tau,\tau''} \qquad (11.31)$$

Essentially what is happening in equation (11.31) is that the functions $\chi^*_{n'l'm'}(\mathbf{x}-\mathbf{X}_{a'})$ and $\chi_{nlm}(\mathbf{x}-\mathbf{X}_a)$ are translated respectively from the points $\mathbf{X}_{a'}$ and $\mathbf{X}_{a''}$ to the point $\mathbf{X}_a$. Here the matrix element of the attractive Coulomb potential of nucleus a is evaluated by means of equations (11.29) and (11.11). Notice that in making this interpretation, we must make use of equation (11.27) to derive the translation properties of the complex conjugated function $\chi^*_{n'l'm'}(\mathbf{x}-\mathbf{X}_{a'})$.

11.5 The Sturmian secular equations for an electron moving in a many-center potential

We are now in a position to derive the secular equations for the Sturmian representation of an electron moving in a many-center Coulomb potential. Multiplying (11.7) by a conjugate function in our basis set integrating over d^3x, and making use of equations (11.31) and (11.10), we obtain the Sturmian secular equations

$$\sum_\tau \left[\mathfrak{W}_{\tau',\tau} - k_j \mathfrak{S}_{\tau',\tau} \right] C_{\tau,j} = 0 \qquad (11.32)$$

The secular equations (11.32) can be written in a different form by introducing the matrix

$$K_{\tau',\tau} \equiv \sqrt{\frac{Z_{a'} Z_a}{n'n}} \mathfrak{S}_{\tau',\tau} \qquad (11.33)$$

Combining (11.33) and (11.31) we have

$$\mathfrak{W}_{\tau',\tau} = \sqrt{\frac{n'n}{Z_{a'} Z_a}} \sum_{\tau''} K_{\tau',\tau''} K_{\tau'',\tau} \qquad (11.34)$$

so that (11.32) becomes

$$\sum_{\tau}\left[\sum_{\tau''} K_{\tau',\tau''} K_{\tau'',\tau} - k_j K_{\tau',\tau}\right] C_{\tau,j} = 0 \qquad (11.35)$$

Now suppose that we have found a set of coefficients $C_{\tau,j}$ that satisfy the secular equations

$$\sum_{\tau} [K_{\tau',\tau} - k_j \delta_{\tau',\tau}] C_{\tau,j} = 0 \qquad (11.36)$$

Then by substituting (11.36) into (11.35), we can see that these coefficients will also be solutions of (11.35). Thus if the basis set is complete, (11.32) and (11.36) contain the same information. However, when the basis set is truncated, as it always is in practice, the information contained in these equations is not precisely the same, and experience has shown that for a small basis, (11.32) gives the most accurate results. This is especially true when direct methods are available for evaluating the nuclear attraction integrals, as is the case for diatomic molecules. However, as more and more Sturmians are added to the basis set, over-completeness begins to be a problem when (11.32) is used, and with a very large basis, (11.36) yields the best results.

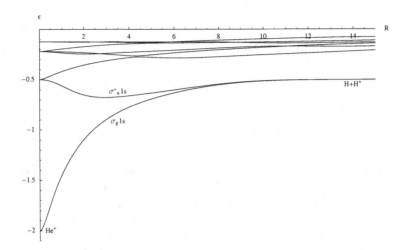

Fig. 11.1 The electron energies of the ground state and excited states of H_2^+ as functions of the internuclear distance, R. The figure was constructed using equation (11.36) with Shibuya-Wulfman integrals generated by means of A. Caligiana's FORTRAN program. In the united-atom limit ($R=0$) the energies correspond to states of He^+, i.e. $\epsilon = -2^2/(2n^2)$, with $n = 1, 2, 3, \ldots$. In the separated-atom limit on the right-hand side of the figure, the energies approach those of the H^++H system, i.e. $\epsilon = -1/(2n^2)$. Only states that are symmetric with respect to rotation about the internuclear axis (σ-states) are shown.

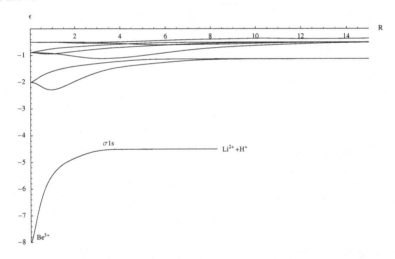

Fig. 11.2 This figure is similar to Figure 11.1, except that the charges on the two nuclei are $Z = 3$ and $Z = 1$ respectively. In the united-atom limit the electronic energies are those of Be^{3+}, i.e. $\epsilon = -4^2/(2n^2)$, while in the separated-atom limit, the lowest energies approach those of the $Li^{2+}+H^+$ system, i.e., $\epsilon = -3^2/(2n^2)$. Only σ-states are shown.

11.6 Molecular spectra

It might be asked whether solutions to the many-center one-electron wave equation can be used to study the spectra of molecules and molecular ions with several electrons. To do this, we can try to construct generalized Sturmians analogous to those used to study the spectra of atoms and atomic ions. Let us suppose that we have solved equation (11.1) with a weighted potential. In other words, suppose that we have solved

$$\left[-\frac{1}{2}\nabla_i^2 + \beta_\nu v(\mathbf{x}_i) - \epsilon_j\right]\varphi_j(\mathbf{x}_i) = 0 \qquad i = 1, 2, ..., N \qquad (11.37)$$

For an N-electron system, we choose the weighting factors β_ν in such a way that

$$\epsilon_1 + \epsilon_2 + ... + \epsilon_N = E_\kappa \qquad (11.38)$$

Then, if we construct Slater determinants of the form

$$\Phi_\nu(\mathbf{x}) = |\varphi_1\varphi_2\varphi_3...| \qquad (11.39)$$

they will be solutions to

$$\left[-\frac{1}{2}\Delta + \beta_\nu V_0(\mathbf{x}) - E_\kappa\right]\Phi_\nu(\mathbf{x}) = 0 \qquad (11.40)$$

where

$$\Delta \equiv \nabla_1^2 + \nabla_2^2 + ... + \nabla_N^2 \qquad (11.41)$$

and

$$V_0(\mathbf{x}) \equiv \sum_{i=1}^{N} v(\mathbf{x}_i) \qquad (11.42)$$

The generalized Sturmian basis functions $\Phi_\nu(\mathbf{x})$ will obey a potential-weighted orthonormality relation, and we can normalize them in such a way that they obey

$$\int d\mathbf{x}\ \Phi_{\nu'}^*(\mathbf{x})V_0(\mathbf{x})\Phi_\nu(\mathbf{x}) = \delta_{\nu',\nu}\frac{2E_\kappa}{\beta_\nu} \qquad (11.43)$$

We can then use the generalized Sturmian basis set to try to solve the true Schrödinger equation

$$\left[-\frac{1}{2}\Delta + V(\mathbf{x}) - E_\kappa\right]\Phi_\nu(\mathbf{x}) = 0 \qquad (11.44)$$

where
$$V(\mathbf{x}) = V_0(\mathbf{x}) + V'(\mathbf{x}) \tag{11.45}$$

and where V' is the interelectron repulsion term

$$V'(\mathbf{x}) = \sum_{i'>i}^{N} \frac{1}{|\mathbf{x}_i - \mathbf{x}_{i'}|} \tag{11.46}$$

If we let

$$\Psi_\kappa(\mathbf{x}) = \sum_\nu \Phi_\nu(\mathbf{x}) B_{\nu,\kappa} \tag{11.47}$$

we obtain (as in Chapter 1) the set of secular equations

$$\sum_\nu \int d x \; \Phi_{\nu'}^*(\mathbf{x}) \left[V(\mathbf{x}) - \beta_\nu V_0(\mathbf{x}) \right] \Phi_\nu(\mathbf{x}) B_{\nu,\kappa} = 0 \tag{11.48}$$

With the help of the potential-weighted orthonormality relations and the definitions

$$T_{\nu',\nu} \equiv -\frac{1}{p_\kappa} \int d x \; \Phi_{\nu'}^*(\mathbf{x}) V(\mathbf{x}) \Phi_\nu(\mathbf{x}) \tag{11.49}$$

and

$$p_\kappa \equiv \sqrt{-2 E_\kappa} \tag{11.50}$$

these secular equations can be written in the familiar form

$$\sum_\nu \left[T_{\nu',\nu} - p_\kappa \delta_{\nu',\nu} \right] B_{\nu,\kappa} = 0 \tag{11.51}$$

Just as in the case of atoms, the nuclear attraction term is diagonal:

$$T_{\nu',\nu}^{(0)} \equiv -\frac{1}{p_\kappa} \int d x \; \Phi_{\nu'}^*(\mathbf{x}) V_0(\mathbf{x}) \Phi_\nu(\mathbf{x}) = \delta_{\nu',\nu} \frac{p_\kappa}{\beta_\nu} \tag{11.52}$$

For a given configuration Φ_ν and a given energy E_κ, we can find the appropriate weighting factors β_ν in the following way:

(1) Solve equation (11.37) for many values of β_ν and make a table of the ϵ_j values corresponding to the solutions $\varphi_j(\mathbf{x}_i)$
(2) For each configuration $\Phi_\nu(\mathbf{x})$, calculate the value of E_κ for every value of β_ν.
(3) Use interpolation to invert the relationship and find β_ν as a function of E_κ.

(4) Solve (11.51) using $p_\kappa = \sqrt{-2E_\kappa}$ for the nuclear attraction term shown in equation (11.52) and find an improved value of p_κ. If necessary, solve (11.51) again using the improved p_κ/β_ν value for the nuclear attraction term.

Equation (11.37) may be solved in a variety of ways. If Shibuya-Wulfman integrals are used to solve it, the solutions will be expressed in terms of basis functions of the type shown in equation (11.5). Many-center interelectron repulsion integrals using exponential-type orbitals (ETO's) such as those in (11.5) are more difficult to evaluate than the corresponding integrals involving Gaussian basis functions. Nevertheless, program packages are available for evaluating interelectron repulsion integrals involving ETO's. Alternatively the methods of Appendix C and Chapter 9 may be used.

A second approach is to use a Gaussian basis both for solving equation (11.37) and for evaluating the many-center interelectron repulsion integrals. In fact, a standard Gaussian program may be modified so as to produce generalized Sturmians for the study of molecular spectra. We hope that future research in this direction will yield interesting results.

Finally, a third Sturmian approach to molecular spectra is possible. We discuss this approach in Appendix F, where generalized Shibuya-Wulfman integrals are used to build up molecular wave functions from atomic configurations based on Coulomb Sturmians.

Table 11.1: The first few Shibuya-Wulfman integrals, $\mathfrak{S}_{\tau',\tau}$, with $\tau \equiv (\alpha, a)$ and $\mathbf{s} \equiv k(\mathbf{X}_{a'} - \mathbf{X}_a)$ with $s \equiv |\mathbf{s}|$. The integrals refer to the real Coulomb Sturmians.

$$\chi_{1s} = \sqrt{k^3/\pi}\ e^{-kr}$$
$$\chi_{2s} = \sqrt{k^3/\pi}\ (1-kr)\ e^{-kr}$$
$$\chi_{2p_j} = \sqrt{k^3/\pi}\ kx_j\ e^{-kr}$$

α'	$\alpha = 1s$	$\alpha = 2s$
$1s$	$(1+s)e^{-s}$	$-2s^2 e^{-s}/3$
$2s$	$-2s^2 e^{-s}/3$	$(3+3s-2s^2+s^3)e^{-s}/3$
$2p_j$	$2s_j(1+s)e^{-s}/3$	$s_j(1+s-s^2)e^{-s}/3$

α'	$\alpha = 2p_1$
$2p_1$	$(3+3s+s^2-s_1^2-ss_1^2)e^{-s}/3$
$2p_2$	$s_1 s_2(1+s)e^{-s}/3$
$2p_3$	$s_1 s_3(1+s)e^{-s}/3$

Table 11.2: Ground-state electronic energies of H_2^+ in Hartrees. The internuclear distance R is measured in Bohrs. The results obtained using equation (11.32) with 3 basis functions on each center are shown in column a. T. Koga and T. Matsuhashi have performed the same calculation with a large number of basis functions [Koga and Matsuhashi, 1988]. Their nearly exact results are shown in column b, compared with the best results obtained using ellipsoidal coordinates (column c).

R	a	b	c	R	a	b	c
0.1	−1.9782	−1.97824014	−1.9782421	2.0	−1.1018	−1.102634214	−1.102634214
0.2	−1.9285	−1.92862027	−1.9286202	3.0	−0.9100	−0.910896197	−0.910896197
0.4	−1.8001	−1.800754051	−1.8007539	4.0	−0.7948	−0.796084884	−0.796084884
0.6	−1.6703	−1.671484711	−1.6714846	6.0	−0.6776	−0.678635715	−0.678635715
0.8	−1.5531	−1.554480093	−1.5544801	8.0	−0.6272	−0.627570389	−0.627570389
1.0	−1.4503	−1.451786313	−1.451786313				

Appendix A

INTERELECTRON REPULSION INTEGRALS

A.1 Procedure for evaluating the interelectron repulsion matrix

The interelectron repulsion matrix has the form (3.32)

$$T'_{\nu',\nu} \equiv -\frac{1}{p_\kappa} \int d\tau\, \Phi^*_{\nu'}(\mathbf{x}) V'(\mathbf{x}) \Phi_\nu(\mathbf{x})$$

where

$$V' = \sum_{i=1}^{N} \sum_{j>i}^{N} \frac{1}{r_{ij}}$$

is a two-electron operator. Since radial orthonormality of atomic orbitals belonging to different configurations cannot be assumed, we need to use the generalized Slater-Condon rule (B.19)

$$\langle \Phi_\nu | V' | \Phi_{\nu'} \rangle = \sum_{i=1}^{N} \sum_{j=i+1}^{N} \sum_{k=1}^{N} \sum_{l=k+1}^{N} (-1)^{i+j+k+l} C_{ij;kl} |S_{ij;kl}|$$

where, from (B.20),

$$C_{ij;kl} \equiv \int d\tau_1 \int d\tau_2\, f_i^*(1) f_j^*(2) \frac{1}{r_{12}} [g_k(1) g_l(2) - g_l(1) g_k(2)]$$

The quantities f and g that appear here are one-electron atomic orbitals of the form defined in equations (3.10)-(3.12). These are polynomials in r_j, multiplied by an exponential function and a spherical harmonic. If we focus on the powers $r_1^{j_1} r_2^{j_2}$ appearing in the products of atomic orbitals, the integrals to be evaluated have the form shown in equation (A.1).

In the relativistic case, the powers of r_j are in general not integers, but nevertheless, the equations of Section A.2 can be used. In the relativistic case, we wish to evaluate matrix elements of the Gaunt operator (7.71), so that $C_{ij;kl}$ takes the form (7.77)

$$C_{ij;kl} \equiv -\frac{1}{c^2} \int d^3x_1 \int d^3x_2 \sum_{\lambda=1}^{4} \frac{\left[j_\lambda^{ij}(\mathbf{x}_1)j_\lambda^{kl}(\mathbf{x}_2) - j_\lambda^{ik}(\mathbf{x}_1)j_\lambda^{jl}(\mathbf{x}_2)\right]}{|\mathbf{x}_1 - \mathbf{x}_2|}$$

where, as defined in equation (7.78),

$$j_\lambda^{\mu',\mu}(\mathbf{x}_1) = ic(\chi_{\mu'}^\dagger(\mathbf{x}_1)(\gamma_0\gamma_\lambda)_1\chi_\mu(\mathbf{x}_1))$$

Because of the relationship (7.86)

$$T'_{\nu',\nu} \equiv -\frac{1}{p_\kappa}V'_{\nu',\nu}$$

we must divide $V'_{\nu'\nu}$ by p_κ to obtain $T'_{\nu'\nu}$. The one-electron Dirac spinors χ_μ that are needed for evaluation of $C_{ij;kl}$ are defined by equations (7.29) through (7.42).

A.2 Separation of the integrals into radial and angular parts

In order to evaluate the interelectron repulsion matrix we need to be able to evaluate integrals of the form

$$J = \int_0^\infty dr_1 \, r_1^{2+j_1} e^{-\zeta_1 r_1} \int_0^\infty dr_2 \, r_2^{2+j_2} e^{-\zeta_2 r_2}$$
$$\times \int d\Omega_1 \, W_1(\hat{\mathbf{x}}_1) \int d\Omega_2 \, W_2(\hat{\mathbf{x}}_2) \frac{1}{r_{12}} \quad (A.1)$$

Expanding $1/r_{12}$ in terms of Legendre polynomials, we have

$$\frac{1}{r_{12}} \equiv \frac{1}{|\mathbf{x}_1 - \mathbf{x}_2|} = \sum_{l=0}^{\infty} \frac{r_<^l}{r_>^{l+1}} P_l(\hat{\mathbf{x}}_1 \cdot \hat{\mathbf{x}}_2) \quad (A.2)$$

so that

$$J = \sum_{l=0}^{\infty} a_l I_l \quad (A.3)$$

where

$$a_l \equiv \int d\Omega_1 \, W_1(\hat{\mathbf{x}}_1) \int d\Omega_2 \, W_2(\hat{\mathbf{x}}_2) P_l(\hat{\mathbf{x}}_1 \cdot \hat{\mathbf{x}}_2) \quad (A.4)$$

and
$$I_l \equiv \int_0^\infty dr_1\, r_1^{j_1+2} e^{-\zeta_1 r_1} \int_0^\infty dr_2\, r_2^{j_2+2} e^{-\zeta_2 r_2} \frac{r_<^l}{r_>^{l+1}} \quad (A.5)$$

A.3 Evaluation of the radial integrals in terms of hypergeometric functions

Since $r_<$ is defined to be the smaller of r_1 or r_2, while $r_>$ is the larger of the two, the radial integral over r_2 in equation (A.5) can be rewritten in the form

$$\int_0^\infty dr_2\, r_2^{j_2+2} e^{-\zeta_2 r_2} \frac{r_<^l}{r_>^{l+1}}$$
$$= \frac{1}{r_1^{l+1}} \int_0^{r_1} dr_2\, r_2^{j_2+l+2} e^{-\zeta_2 r_2} + r_1^l \int_{r_1}^\infty dr_2\, r_2^{j_2-l+1} e^{-\zeta_2 r_2} \quad (A.6)$$

The two integrals in equation (A.6) can be expressed in terms of incomplete gamma functions. The first integral can be evaluated in the form

$$\frac{1}{r_1^{l+1}} \int_0^{r_1} dr_2\, r_2^{j_2+l+2} e^{-\zeta_2 r_2} = \frac{1}{(\zeta_2 r_1)^{l+1} \zeta_2^{j_2+2}} \gamma(j_2+l+3, \zeta_2 r_1) \quad (A.7)$$

while the second is

$$r_1^l \int_{r_1}^\infty dr_2\, r_2^{j_2-l+1} e^{-\zeta_2 r_2} = \frac{(\zeta_2 r_1)^l}{\zeta_2^{j_2+2}} \Gamma(j_2-l+2, \zeta_2 r_1) \quad (A.8)$$

where

$$\gamma(a,z) \equiv \int_0^z dt\, t^{a-1} e^{-t} \quad (A.9)$$

and

$$\Gamma(a,z) \equiv \int_z^\infty dt\, t^{a-1} e^{-t} \quad (A.10)$$

The integration over r_1 can also be evaluated analytically, and the final result can be expressed in terms of hypergeometric functions[2]. Thus finally we obtain the double radial integral of equation (A.5) in the form:

$$I_l = \int_0^\infty dr_1\, r_1^{j_1+2} e^{-\zeta_1 r_1} \int_0^\infty dr_2\, r_2^{j_2+2} e^{-\zeta_2 r_2} \frac{r_<^l}{r_>^{l+1}}$$

---
[2]See for example [Gradshteyn and Ryshik, 1965].

$$= \frac{\Gamma(j_1+j_2+5)}{(\zeta_1+\zeta_2)^{j_1+j_2+5}}\left[\frac{{}_2F_1\left(1,j_1+j_2+5|j_2+l+4|\zeta_2/(\zeta_1+\zeta_2)\right)}{j_2+l+3}\right.$$
$$\left.+\frac{{}_2F_1\left(1,j_1+j_2+5|j_1+l+4|\zeta_1/(\zeta_1+\zeta_2)\right)}{j_1+l+3}\right] \quad \text{(A.11)}$$

In the non-relativistic case, j_1 and j_2 are integers, but in the relativistic case they are non-integral. However, even when j_1 and j_2 are non-integral, Mathematica is able to evaluate the hypergeometric functions in (A.11) very rapidly.

A.4 Evaluation of the angular integrals by harmonic projection

Let us now turn to the angular integrals of equation (A.4). In all cases of interest to us, W_1 and W_2 can be expressed as homogeneous polynomials in the components of the unit vectors $\hat{\mathbf{x}}_1$ and $\hat{\mathbf{x}}_2$. Thus, we can use the properties of harmonic polynomials (discussed in Chapters 9 and 10) to carry out the angular integrations. We can remember from Chapters 9 and 10 that

$$\int d\Omega_2\; W_2(\hat{\mathbf{x}}_2) P_l(\hat{\mathbf{x}}_1\cdot\hat{\mathbf{x}}_2) = \frac{4\pi}{2l+1} O_l[W_2(\hat{\mathbf{x}}_1)] \quad \text{(A.12)}$$

where O_l is a harmonic projection operator that projects out the harmonic component of W_2 corresponding to the angular momentum quantum number l. Equation (A.12) is a consequence of the sum rule (10.53). We can also remember from Chapters 9 and 10 that

$$\int d\Omega_1\; W_1(\hat{\mathbf{x}}_1) O_l[W_2(\hat{\mathbf{x}}_1)] = 4\pi O_0\left[W_1(\hat{\mathbf{x}}) O_l[W_2(\hat{\mathbf{x}})]\right] \quad \text{(A.13)}$$

Thus we obtain the relationship

$$a_l = \frac{(4\pi)^2}{2l+1} O_0\left[W_1(\hat{\mathbf{x}}) O_l[W_2(\hat{\mathbf{x}})]\right] \quad \text{(A.14)}$$

which can also be written in the more symmetrical form

$$a_l = \frac{(4\pi)^2}{2l+1} O_0\left[O_l[W_1(\hat{\mathbf{x}})] O_l[W_2(\hat{\mathbf{x}})]\right] \quad \text{(A.15)}$$

In equation (A.15), O_0 projects out the constant term in the harmonic decomposition of the homogeneous polynomial $O_l[W_1(\hat{\mathbf{x}})]O_l[W_2(\hat{\mathbf{x}})]$. If we let

$$\hat{\mathbf{x}} = (u_1, u_2, u_3) \quad \text{(A.16)}$$

then $W_1(\hat{\mathbf{x}})$ and $W_2(\hat{\mathbf{x}})$ will be homogeneous polynomials in u_1, u_2 and u_3. Suppose that the degrees of these polynomials are n and n', where $n \leq n'$. If n and n' do not have the same parity, then a_l will be zero for all values of l. When n and n' are both odd or both even, non-zero values of a_l may occur for $l = n, n-2, n-4, \ldots$. Thus we can write

$$J = \sum_{l=n,n-2,\ldots} a_l I_l \qquad (A.17)$$

The lowest value of l for which a_l is non-zero is either 0 or 1, depending on whether n is even or odd. As was discussed in Chapters 9 and 10, the harmonic projections O_l and O_0 may be performed by successive applications of the Laplacian operator

$$\nabla_u^2 \equiv \sum_{j=1}^{3} \frac{\partial^2}{\partial u_j^2} \qquad (A.18)$$

If f_n is a homogeneous polynomial of degree n in u_1, u_2 and u_3, and if l has the same parity as n, with $l \leq n$, then from equation (9.32) with $d = 3$ and $\lambda = l$ we have:

$$O_l[f_n] = \frac{(2l+1)!!}{(n-l)!!(n+l+1)!!}$$
$$\times \sum_{j=0}^{[l/2]} \frac{(-1)^j(2l-2j-1)!!}{(2j)!!(2l-1)!!}(u_1^2+u_2^2+u_3^2)^{2j}(\nabla_u^2)^{j+(n-l)/2} f_n$$
$$\qquad (A.19)$$

If n and l have different parities, or if $l > n$, then $O_l[f_n] = 0$. From the discussion in Chapters 9 and 10 it also follows that if f_{2l} is a homogeneous polynomial of degree $2l$ in u_1, u_2 and u_3, then

$$O_0[f_{2l}] = \frac{1}{(2l+1)!}(\nabla_u^2)^l f_{2l} \qquad (A.20)$$

In the non-relativistic case the angular integrals a_l can alternatively be evaluated by means of Clebsch-Gordan coefficients, and with sufficient effort this can also be done in the relativistic case. However, in the relativistic case, the method of harmonic projection outlined above is by far the most convenient.

A.5 Relativistic interelectron repulsion integrals

In Chapter 4, we discussed the 4-component solutions to the Dirac equation for hydrogenlike atoms. These have the form

$$\chi_{n,j,l,M}(\mathbf{x}) = \begin{pmatrix} ig_{njl}(r)\Omega_{j,l,M}(\mathbf{u}) \\ -f_{njl}(r)\Omega_{j,2j-l,M}(\mathbf{u}) \end{pmatrix} \quad (A.21)$$

where $\Omega_{j,l,M}(\mathbf{u})$ is a 2-component spherical spinor. If we let μ stand for the set of indices n, j, l and m, and if we introduce the notation

$$\tilde{\Omega}_{j,l,M}(\mathbf{u}) \equiv \Omega_{j,2j-l,M}(\mathbf{u}) \quad (A.22)$$

then (A.21) can be rewritten in the more condensed form

$$\chi_\mu(\mathbf{x}) = \begin{pmatrix} ig_\mu(r)\Omega_\mu(\mathbf{u}) \\ -f_\mu(r)\tilde{\Omega}_\mu(\mathbf{u}) \end{pmatrix} \quad (A.23)$$

while $\chi^\dagger_{\mu'}(\mathbf{x})$ becomes

$$\chi^\dagger_{\mu'}(\mathbf{x}) = \left(-ig_{\mu'}(r)\Omega^\dagger_{\mu'}(\mathbf{u}),\; -f_{\mu'}(r)\tilde{\Omega}^\dagger_{\mu'}(\mathbf{u})\right) \quad (A.24)$$

In this notation, the transition density becomes

$$\rho_{\mu',\mu} = (\chi^\dagger_{\mu'}\chi_\mu) = g_{\mu'}g_\mu(\Omega^\dagger_{\mu'}\Omega_\mu) + f_{\mu'}f_\mu(\tilde{\Omega}^\dagger_{\mu'}\tilde{\Omega}_\mu) \quad (A.25)$$

Similarly, in our condensed notation, the transition current vector becomes

$$\mathbf{j}_{\mu',\mu} = c(\chi^\dagger_{\mu'}\boldsymbol{\alpha}\chi_\mu) = icg_{\mu'}f_\mu(\Omega^\dagger_{\mu'}\boldsymbol{\sigma}\tilde{\Omega}_\mu) - icf_{\mu'}g_\mu(\tilde{\Omega}^\dagger_{\mu'}\boldsymbol{\sigma}\Omega_\mu) \quad (A.26)$$

where $\boldsymbol{\sigma}$ is a vector whose components are the Pauli spin matrices. We need to evaluate integrals of the form

$$I = -\frac{1}{c^2}\int d^3x_1 \int d^3x_2 \sum_{\lambda=1}^{4} \frac{j_\lambda^{\mu''',\mu''}(\mathbf{x}_1)j_\lambda^{\mu',\mu}(\mathbf{x}_2)}{|\mathbf{x}_1 - \mathbf{x}_2|} \quad (A.27)$$

Expanding the denominator in terms of Legendre polynomials, we have then

$$I = -\frac{1}{c^2}\sum_l \int_0^\infty dr_1\, r_1^2 \int_0^\infty dr_2\, r_2^2 \frac{r_<^l}{r_>^{l+1}} F_l(r_1,r_2) \quad (A.28)$$

where

$$F_l(r_1, r_2) = \frac{(4\pi)^2}{2l+1} \sum_{\lambda=1}^{4} O_0 \left[O_l[j_\lambda^{\mu''',\mu''}(r_1, \mathbf{u})] O_l[j_\lambda^{\mu',\mu}(r_2, \mathbf{u})] \right] \quad \text{(A.29)}$$

If we introduce the 2-component vectors

$$\mathbf{w}_\lambda^{\mu',\mu}(\mathbf{u}) \equiv \begin{cases} \left((\Omega_{\mu'}^\dagger \sigma_\lambda \tilde{\Omega}_\mu), (\tilde{\Omega}_{\mu'}^\dagger \sigma_\lambda \Omega_\mu) \right) & \lambda = 1, 2, 3 \\ \left((\Omega_{\mu'}^\dagger \Omega_\mu), (\tilde{\Omega}_{\mu'}^\dagger \tilde{\Omega}_\mu) \right) & \lambda = 4 \end{cases} \quad \text{(A.30)}$$

and

$$\mathbf{v}_\lambda^{\mu',\mu}(r_1) \equiv \begin{cases} \left(ig_{\mu'}(r_1) f_\mu(r_1), -i f_{\mu'}(r_1) g_\mu(r_1) \right) & \lambda = 1, 2, 3 \\ \left(ig_{\mu'}(r_1) g_\mu(r_1), i f_{\mu'}(r_1) f_\mu(r_1) \right) & \lambda = 4 \end{cases} \quad \text{(A.31)}$$

then we can express the 4-currents in the form

$$j_\lambda^{\mu',\mu}(r_1, \mathbf{u}) = c\, \mathbf{v}_\lambda^{\mu',\mu}(r_1) \cdot \mathbf{w}_\lambda^{\mu',\mu}(\mathbf{u}) \quad \text{(A.32)}$$

Substituting this expression into (A.29), we obtain

$$F_l(r_1, r_2) = \frac{(4\pi c)^2}{2l+1} \sum_{\lambda=1}^{4} O_0 \left[[\mathbf{v}_\lambda^{\mu''',\mu''}(r_1) \cdot \mathbf{w}_{\lambda,l}^{\mu''',\mu''}(\mathbf{u})] [\mathbf{v}_\lambda^{\mu',\mu}(r_2) \cdot \mathbf{w}_{\lambda,l}^{\mu',\mu}(\mathbf{u})] \right] \quad \text{(A.33)}$$

where

$$\mathbf{w}_{\lambda,l}^{\mu',\mu}(\mathbf{u}) \equiv O_l[\mathbf{w}_\lambda^{\mu',\mu}(\mathbf{u})] \quad \text{(A.34)}$$

Appendix B

GENERALIZED SLATER-CONDON RULES

B.1 Introduction

The Slater-Condon rules allow us to evaluate the matrix elements of 1-electron and 2-electron operators between N-electron configurations. In the usual derivation of these rules it is assumed that these configurations are Slater determinants constructed from an orthonormal set of 1-electron spin-orbitals. However, in the generalized Sturmian method, the spin-orbitals of one configuration cannot be assumed to be orthogonal to the spin-orbitals of another configuration. For example, if Goscinskian configurations are used, each configuration is characterized by its own value of effective nuclear charge, Q_ν, and thus radial orthonormality is lost.

Generalized Slater-Condon rules (i.e. rules for constructing matrix elements between Slater determinants constructed from spin-orbitals that are not necessarily orthonormal) have been discussed by a number of authors [Amos and Hall, 1961], [Löwdin, 1962], [King et al., 1967], [McWeeny, 1989]. After trying several algorithms, we have found that the version of the generalized Slater-Condon rules described by Rettrup is the most robust and computationally stable [Rettrup, 2003].

B.2 Slater determinants expressed in terms of the antisymmetrizer

To derive the generalized Slater-Condon rules in Rettrup's form, we introduce the antisymmetrizer

$$A_N = \sum_P \epsilon_P P \qquad (B.1)$$

Here P permutes the electron coordinates and ϵ_P is the sign associated with the permutation. There are $N!$ terms in the sum shown in equation (B.1). When it is normalized in the manner shown in (B.1), the antisymmetrizer obeys the relationship

$$A_N^2 = N! A_N \tag{B.2}$$

A Slater determinant, Φ_ν, can be expressed in terms of the antisymmetrizer acting on a product of spin-orbitals:

$$\sqrt{N!}\Phi_\nu \equiv F \equiv \begin{vmatrix} f_1(1) & f_2(1) & \cdots & f_N(1) \\ f_1(2) & f_2(2) & \cdots & f_N(2) \\ \vdots & \vdots & & \vdots \\ f_1(N) & f_2(N) & \cdots & f_N(N) \end{vmatrix} = A_N f_1(1) f_2(2)...f_N(N) \tag{B.3}$$

Similarly,

$$\sqrt{N!}\Phi_{\nu'} \equiv G \equiv \begin{vmatrix} g_1(1) & g_2(1) & \cdots & g_N(1) \\ g_1(2) & g_2(2) & \cdots & g_N(2) \\ \vdots & \vdots & & \vdots \\ g_1(N) & g_2(N) & \cdots & g_N(N) \end{vmatrix} = A_N g_1(1) g_2(2)...g_N(N) \tag{B.4}$$

B.3 Scalar products between configurations

The scalar product between two configurations can then be expressed in the form

$$\begin{aligned} \langle \Phi_\nu | \Phi_{\nu'} \rangle &= \frac{1}{N!} \langle F | G \rangle \\ &= \frac{1}{N!} \int d\tau_1 \cdots \int d\tau_N \, (A_N f_1(1) \cdots f_N(N))^* (A_N g_1(1) \cdots g_N(N)) \\ &= \int d\tau_1 \cdots \int d\tau_N \, f_1^*(1) \cdots f_N^*(N) \, (A_N g_1(1) \cdots g_N(N)) \end{aligned}$$

SLATER-CONDON RULES

$$= \int d\tau_1 \cdots \int d\tau_N \, f_1^*(1) \cdots f_N^*(N) \begin{vmatrix} g_1(1) & g_2(1) & \cdots & g_N(1) \\ g_1(2) & g_2(2) & \cdots & g_N(2) \\ \vdots & \vdots & & \vdots \\ g_1(N) & g_2(N) & \cdots & g_N(N) \end{vmatrix}$$
(B.5)

where we have made use of the relationship (B.2). Carrying out the integrations in (B.5), we obtain the first generalized Slater-Condon rule:

$$\langle \Phi_\nu | \Phi_{\nu'} \rangle = |S| \tag{B.6}$$

where S is the matrix of overlap integrals:

$$S \equiv \begin{pmatrix} \langle f_1|g_1 \rangle & \langle f_1|g_2 \rangle & \cdots & \langle f_1|g_N \rangle \\ \langle f_2|g_1 \rangle & \langle f_2|g_2 \rangle & \cdots & \langle f_2|g_N \rangle \\ \vdots & \vdots & & \vdots \\ \langle f_3|g_1 \rangle & \langle f_3|g_2 \rangle & \cdots & \langle f_3|g_N \rangle \end{pmatrix} \tag{B.7}$$

B.4 One-electron operators

We next consider matrix elements of one-electron operators of the form

$$V = v(1) + v(2) + \cdots + v(N) \tag{B.8}$$

From (B.3) and (B.4) and from the fact that each term in the sum shown in (B.8) gives the same contribution, we have

$$\langle \Phi_\nu | V | \Phi_{\nu'} \rangle = \frac{1}{N!} \langle F|V|G \rangle = \frac{N}{N!} \langle F|v(1)|G \rangle \tag{B.9}$$

The determinant F of equation (B.3) can be expanded in terms of its minors:

$$F = \sum_{i=1}^{N} (-1)^{i+1} f_i(1) F_i^{N-1} \tag{B.10}$$

Here F_i^{N-1} is the matrix obtained from F by deleting the first row and the ith column. A similar expansion can be made for G:

$$G = \sum_{j=1}^{N}(-1)^{j+1}g_j(1)G_j^{N-1} \tag{B.11}$$

With the help of these two expansions, we can write:

$$\langle F|v(1)|G\rangle = \sum_{i=1}^{N}\sum_{j=1}^{N}(-1)^{i+j}\langle f_i|v|g_j\rangle\langle F_i^{N-1}|G_j^{N-1}\rangle \tag{B.12}$$

where

$$\langle f_i|v|g_j\rangle \equiv \int d\tau_1\, f_i^*(1)v(1)g_j(1) \tag{B.13}$$

Using the methods of equations (B.3)-(B.7) to evaluate $\langle F_i^{N-1}|G_j^{N-1}\rangle$, we obtain the second generalized Slater-Condon rule:

$$\langle \Phi_\nu|V|\Phi_{\nu'}\rangle = \sum_{i=1}^{N}\sum_{j=1}^{N}(-1)^{i+j}\langle f_i|v|g_j\rangle|S_{ij}| \tag{B.14}$$

where $|S_{ij}|$ is the determinant obtained from $|S|$ by deleting the ith row and the jth column.

B.5 Two-electron operators

Finally, let us consider 2-electron operators, such as the interelectron repulsion operator:

$$V' = \sum_{i>j}^{N}\sum_{j=1}^{N}\frac{1}{r_{ij}} \tag{B.15}$$

Since each term in the double sum gives the same contribution, and since there are $N(N-1)/2$ terms, we can write

$$\langle \Phi_\nu|V'|\Phi_{n u'}\rangle = \frac{N(N-1)}{2}\langle \Phi_\nu|\frac{1}{r_{12}}|\Phi_{\nu'}\rangle = \frac{1}{2(N-2)!}\langle F|\frac{1}{r_{12}}|G\rangle \tag{B.16}$$

where we have also made use of equations (B.3) and (B.4). We next expand the determinants F and G in terms of their double minors:

$$F = \sum_{i=1}^{N}\sum_{j=i+1}^{N}(-1)^{i+j+1}[f_i(1)f_j(2) - f_j(1)f_i(2)]F_{ij}^{N-2} \tag{B.17}$$

Here F_{ij}^{N-2} is obtained from F by deleting the first and second rows and the ith and jth columns. Similarly,

$$G = \sum_{k=1}^{N} \sum_{l=k+1}^{N} (-1)^{k+l+1} \left[g_k(1)g_l(2) - g_l(1)g_k(2) \right] G_{kl}^{N-2} \qquad (B.18)$$

Inserting the expansions (B.17) and (B.18) into (B.16), we obtain the third generalized Slater-Condon rule:

$$\langle \Phi_\nu | V' | \Phi_{\nu'} \rangle = \sum_{i=1}^{N} \sum_{j=i+1}^{N} \sum_{k=1}^{N} \sum_{l=k+1}^{N} (-1)^{i+j+k+l} C_{ij;kl} |S_{ij;kl}| \qquad (B.19)$$

where $|S_{ij;kl}|$ is the determinant of the matrix obtained from S by deleting the ith and jth rows and the kth and lth columns, and where

$$C_{ij;kl} \equiv \int d\tau_1 \int d\tau_2 \; f_i^*(1) f_j^*(2) \frac{1}{r_{12}} \left[g_k(1) g_l(2) - g_l(1) g_k(2) \right] \qquad (B.20)$$

The determinants such as $|S|$, $|S_{ij}|$ and $|S_{ij;kl}|$ consist of numbers rather than algebraic expressions, and modern algorithms for evaluating them are rapid.

Appendix C

EXPANSION OF $F(r)$ ABOUT ANOTHER CENTER

C.1 Expansion of a displaced function of r in terms of Gegenbauer polynomials

Let $F(r)$ be some function of the hyperradius r in a d-dimensional space, where

$$r \equiv \left(x_1^2 + x_2^2 + \ldots + x_d^2\right)^{1/2} \tag{C.1}$$

If we displace the center of our coordinate system by a vector $\mathbf{x}'$, the function becomes $F(|\mathbf{x}-\mathbf{x}'|)$. We would like to show that the displaced function can be expanded in a series of the form

$$F(|\mathbf{x}-\mathbf{x}'|) = \sum_{\lambda=0}^{\infty} f_\lambda(r,r') C_\lambda^\alpha(\mathbf{u} \cdot \mathbf{u}') \tag{C.2}$$

where C_λ^α is a Gegenbauer polynomial and where

$$\alpha \equiv \frac{d}{2} - 1 \tag{C.3}$$

with

$$\mathbf{u} \equiv \frac{1}{r}(x_1, x_2, \ldots, x_d)$$

$$\mathbf{u}' \equiv \frac{1}{r'}(x'_1, x'_2, \ldots, x'_d) \tag{C.4}$$

To make this expansion, we first express $F(r)$ in terms of its d-dimensional Fourier transform

$$F(r) = \frac{1}{(2\pi)^{d/2}} \int dp\, F^t(p)\, e^{i\mathbf{p}\cdot\mathbf{x}} \tag{C.5}$$

where
$$F^t(p) = \frac{1}{(2\pi)^{d/2}} \int dx\, F(r)\, e^{-i\mathbf{p}\cdot\mathbf{x}} \tag{C.6}$$

is a function of the momentum-space hyperradius
$$p \equiv \left(p_1^2 + p_2^2 + \ldots + p_d^2\right)^{1/2} \tag{C.7}$$

alone, i.e., $F^t(p)$ is independent of the momentum-space hyperangles. The displaced function will then be given by
$$F(|\mathbf{x} - \mathbf{x}'|) = \frac{1}{(2\pi)^{d/2}} \int dp\, F^t(p)\, e^{i\mathbf{p}\cdot(\mathbf{x}-\mathbf{x}')} \tag{C.8}$$

C.2 Expansion of plane waves using Gegenbauer polynomials

A d-dimensional plane wave can be expanded in a series of the form
$$e^{i\mathbf{p}\cdot\mathbf{x}} = (d-4)!! \sum_{\lambda=0}^{\infty} i^\lambda (d + 2\lambda - 2) j_\lambda^d(pr) C_\lambda^\alpha(\mathbf{u}_p \cdot \mathbf{u}) \tag{C.9}$$

where the "hyperspherical Bessel function" $j_\lambda^d(pr)$ is related to the familiar Bessel function J by
$$j_\lambda^d(pr) = \frac{\Gamma(\alpha) 2^{\alpha-1} J_{\alpha+\lambda}(pr)}{(d-4)!!(pr)^\alpha} \tag{C.10}$$

(see Appendix E, Chapter 10, and J.S. Avery [Avery, 1989]). Thus we can also write (C.9) in the form
$$e^{i\mathbf{p}\cdot\mathbf{x}} = \frac{\Gamma(\alpha) 2^\alpha}{(pr)^\alpha} \sum_{\lambda=0}^{\infty} i^\lambda (\alpha + \lambda) J_{\alpha+\lambda}(pr) C_\lambda^\alpha(\mathbf{u}_p \cdot \mathbf{u}) \tag{C.11}$$

so that
$$e^{i\mathbf{p}\cdot(\mathbf{x}-\mathbf{x}')} = \frac{(\Gamma(\alpha) 2^\alpha)^2}{(pr)^\alpha (pr')^\alpha} \sum_{\lambda=0}^{\infty} i^\lambda (\alpha + \lambda) J_{\alpha+\lambda}(pr) C_\lambda^\alpha(\mathbf{u}_p \cdot \mathbf{u})$$
$$\times \sum_{\lambda'=0}^{\infty} (-i)^{\lambda'} (\alpha + \lambda') J_{\alpha+\lambda'}(pr') C_{\lambda'}^\alpha(\mathbf{u}_p \cdot \mathbf{u}') \tag{C.12}$$

But the hyperspherical harmonics have a sum rule of the form [Avery, 1989]

$$C_\lambda^\alpha(\mathbf{u}_p \cdot \mathbf{u}) = \frac{\alpha I(0)}{\alpha + \lambda} \sum_\mu Y_{\lambda,\mu}^*(\mathbf{u}_p) Y_{\lambda,\mu}(\mathbf{u}) \qquad (C.13)$$

where $I(0)$ is the generalized total solid angle

$$I(0) \equiv \int d\Omega = \frac{2\pi^{d/2}}{\Gamma(d/2)} = \frac{2\pi^{\alpha+1}}{\Gamma(\alpha+1)} \qquad (C.14)$$

from which it follows that [Avery, 1989]

$$\int d\Omega_p \, C_\lambda^\alpha(\mathbf{u}_p \cdot \mathbf{u}) C_{\lambda'}^\alpha(\mathbf{u}_p \cdot \mathbf{u}') = \delta_{\lambda,\lambda'} \frac{\alpha I(0)}{\alpha + \lambda} C_\lambda^\alpha(\mathbf{u} \cdot \mathbf{u}') \qquad (C.15)$$

Combining (C.15) and (C.12), we obtain:

$$\int d\Omega_p \, e^{i\mathbf{p}\cdot(\mathbf{x}-\mathbf{x}')} = \frac{(\Gamma(\alpha)2^\alpha)^2 \alpha I(0)}{(pr)^\alpha (pr')^\alpha} \sum_{\lambda=0}^\infty (\alpha+\lambda) J_{\alpha+\lambda}(pr) J_{\alpha+\lambda}(pr') C_\lambda^\alpha(\mathbf{u}\cdot\mathbf{u}') \qquad (C.16)$$

C.3 Explicit expressions for the displaced function in terms of integrals over Bessel functions

Finally, substituting (C.16) into (C.8), and remembering that the volume element in momentum space can be expressed in the form

$$dp_1 dp_2 ... dp_d = dp \, p^{d-1} d\Omega_p \qquad (C.17)$$

we obtain

$$F(|\mathbf{x} - \mathbf{x}'|) = \sum_{\lambda=0}^\infty f_\lambda(r,r') C_\lambda^\alpha(\mathbf{u}\cdot\mathbf{u}') \qquad (C.18)$$

where

$$f_\lambda(r,r') = I(0)\alpha(\alpha+\lambda) \frac{(\Gamma(\alpha)2^\alpha)^2}{(2\pi)^{\alpha+1} r^\alpha r'^\alpha} \int_0^\infty dp \, p \, F^t(p) J_{\alpha+\lambda}(pr) J_{\alpha+\lambda}(pr') \qquad (C.19)$$

When $d = 3$, $\alpha \equiv \frac{d}{2} - 1 = \frac{1}{2}$, and (C.18) becomes

$$F(|\mathbf{x} - \mathbf{x}'|) = \sum_{l=0}^\infty f_l(r,r') P_l(\mathbf{u}\cdot\mathbf{u}') \qquad (C.20)$$

where P_l is a Legendre polynomial, while (C.19) becomes

$$f_l(r,r') = (2l+1)\sqrt{\frac{\pi}{2r'r}} \int_0^\infty dp\, p\, F^t(p) J_{\lambda+\frac{1}{2}}(pr) J_{\lambda+\frac{1}{2}}(pr') \qquad \text{(C.21)}$$

Alternatively $f_l(r,r')$ can be expressed in terms of spherical Bessel functions. Since

$$J_{\lambda+\frac{1}{2}}(pr) = \sqrt{\frac{2pr}{\pi}} j_l(pr) \qquad \text{(C.22)}$$

we have

$$f_l(r,r') = (2l+1)\sqrt{\frac{2}{\pi}} \int_0^\infty dp\, p^2\, F^t(p) j_l(pr) j_l(pr') \qquad \text{(C.23)}$$

where (when $d=3$) equation (C.6) reduces to

$$F^t(p) = \sqrt{\frac{2}{\pi}} \int_0^\infty dr\, r^2 F(r) j_l(pr) \qquad \text{(C.24)}$$

C.4 An alternative method, illustrated for the case where $d=3$

A second independent method for expanding a displaced function in terms of Gegenbauer polynomials is discussed in Appendix D of [Avery, 1989]. In this appendix the method is discussed for general d, but we can illustrate it here for $d=3$. If we multiply (C.20) from the left by $P_{l'}(\mathbf{u}\cdot\mathbf{u}')$ and integrate over solid angle, we obtain:

$$\int d\Omega\, P_{l'}(\mathbf{u}\cdot\mathbf{u}') F(|\mathbf{x}-\mathbf{x}'|) = \sum_{l=0}^\infty f_l(r,r') \int d\Omega\, P_{l'}(\mathbf{u}\cdot\mathbf{u}') P_l(\mathbf{u}\cdot\mathbf{u}') \qquad \text{(C.25)}$$

The angular integral on the right-hand side of (C.25) can be performed by means of the sum rule for spherical harmonics, and we obtain

$$f_l(r,r') = \frac{2l+1}{4\pi} \int d\Omega\, P_l(\mathbf{u}\cdot\mathbf{u}') F(|\mathbf{x}-\mathbf{x}'|) \qquad \text{(C.26)}$$

Remembering that

$$|\mathbf{x}-\mathbf{x}'| = \left(r^2 + r'^2 - 2rr'\cos\theta\right)^{1/2} \qquad \text{(C.27)}$$

where

$$\cos\theta \equiv \mathbf{u}\cdot\mathbf{u}' \qquad \text{(C.28)}$$

we can rewrite (C.26) in the form

$$f_l(r,r') = \frac{2l+1}{2} \int_0^\pi d\theta \; \sin\theta P_l(\cos\theta) F\left((r^2 + r'^2 - 2rr'\cos\theta)^{\frac{1}{2}}\right) \quad \text{(C.29)}$$

The integral in (C.29) can be evaluated directly, for example by expanding $F\left((r^2 + r'^2 - 2rr'\cos\theta)^{1/2}\right)$ as a Taylor series in $\cos\theta$. Alternatively we can introduce the variable r'' defined by

$$r'' \equiv \left(r^2 + r'^2 - 2rr'\cos\theta\right)^{1/2} \quad \text{(C.30)}$$

Then

$$\cos\theta = \frac{r^2 + r'^2 - r''^2}{2rr'} \quad \text{(C.31)}$$

and

$$\sin\theta \; d\theta = \frac{r''}{rr'} \; dr'' \quad \text{(C.32)}$$

while equation (C.29) becomes

$$f_l(r,r') = \frac{2l+1}{2rr'} \int_{|r-r'|}^{r+r'} dr'' \; r'' F(r'') P_l\left(\frac{r^2 + r'^2 - r''^2}{2rr'}\right) \quad \text{(C.33)}$$

For example, suppose that

$$F(r) = e^{-\zeta r} \quad \text{(C.34)}$$

and $l = 0$. Then (C.33) yields

$$f_0(r,r') = \frac{1}{2rr'} \int_{|r-r'|}^{r+r'} dr'' \; r'' e^{-\zeta r''}$$

$$= \frac{1}{2\zeta^2 rr'} \left[e^{-\zeta|r-r'|}(1 + \zeta|r-r'|) - e^{-\zeta(r+r')}(1 + \zeta r + \zeta r') \right] \quad \text{(C.35)}$$

The interested reader can verify by means of contour integration that equation (C.23) yields the same result. For the case where

$$F(r) = r^n \; e^{-\zeta r} \quad \text{(C.36)}$$

we have

$$f_0(r,r') = \frac{1}{2rr'} \int_{|r-r'|}^{r+r'} dr'' \; r''^{n+1} e^{-\zeta r''}$$

$$= \frac{1}{2rr'} \left(-\frac{\partial}{\partial \zeta}\right)^{n+1} \int_{|r-r'|}^{r+r'} dr'' \, e^{-\zeta r''}$$
(C.37)

For higher values of l, the function $f_l(r,r')$ is related to $f_0(r,r')$ by

$$f_l(r,r') = (2l+1) P_l(\mathcal{W}) f_0(r,r')$$
(C.38)

where the operator $\mathcal{W}$ is defined as

$$\mathcal{W} \equiv \frac{r^2 + r'^2}{2rr'} - \frac{1}{2rr'} \frac{\partial^2}{\partial \zeta^2}$$
(C.39)

C.5 Closed-form differential expressions in terms of modified spherical Bessel functions

By comparing equations (C.33)-(C.39) with (E.31)-(E.41), we can see that $f_l(r,r')$ can also be written in the form

$$f_l(r,r') = (2l+1)\left(-\frac{\partial}{\partial \zeta}\right)^{n+1} [\zeta \, i_l(\zeta r_<) k_l(\zeta r_>)]$$
(C.40)

where i_l and k_l are respectively modified spherical Bessel functions of the first and second kind.

Appendix D

THE FOCK PROJECTION

D.1 The one-electron Schrödinger equation in momentum space

A one-electron wave function $\chi(\mathbf{x})$ and its Fourier transform $\chi(\mathbf{x})$ are related to each other by

$$\chi(\mathbf{x}) = \frac{1}{(2\pi)^{3/2}} \int d^3p \, e^{i\mathbf{p}\cdot\mathbf{x}} \chi^t(\mathbf{p})$$

$$\chi^t(\mathbf{p}) = \frac{1}{(2\pi)^{3/2}} \int d^3x \, e^{-i\mathbf{p}\cdot\mathbf{x}} \chi(\mathbf{x}) \qquad (D.1)$$

If we introduce a parameter k that is related to the energy by

$$k^2 \equiv -2E \qquad (D.2)$$

then the Schrödinger equation obeyed by $\chi(\mathbf{x})$ can be written in the form

$$\left[-\nabla^2 + k^2 + 2V(\mathbf{x})\right] \chi(\mathbf{x}) = 0 \qquad (D.3)$$

By an argument similar to equations (2.6)-(2.12) we can obtain an integral equation in momentum space that corresponding to the direct-space equation (D.3):

$$(p'^2 + k^2)\chi^t(\mathbf{p}') = -\frac{2}{(2\pi)^{3/2}} \int d^3p \, V^t(\mathbf{p}' - \mathbf{p})\chi^t(\mathbf{p}) \qquad (D.4)$$

where

$$V^t(\mathbf{p}' - \mathbf{p}) \equiv \frac{1}{(2\pi)^{3/2}} \int dx \, e^{-i(\mathbf{p}'-\mathbf{p})\cdot\mathbf{x}} V(\mathbf{x}) \qquad (D.5)$$

Equation (D.4) is the momentum-space Schrödinger equation, expressed in atomic units, for a single electron moving in the potential V.

D.2 The momentum-space wave equation for hydrogenlike atoms

In the case of a hydrogenlike atom

$$V(\mathbf{x}) = -\frac{Z}{r} \tag{D.6}$$

where Z is the nuclear charge. Since the $l = 0$ term in the familiar expansion of a plane wave in terms of spherical Bessel functions and spherical harmonics is just a spherical Bessel function of order zero, we can write the $l = 0$ projection of the plane wave in equation (D.5) as

$$O_0[e^{-i(\mathbf{p}'-\mathbf{p})\cdot\mathbf{x}}] = j_0(|\mathbf{p}' - \mathbf{p}|r) \tag{D.7}$$

Using (D.7), one finds that with the potential $V = -Z/r$, the integral in (D.7) is oscillatory, but with the help of a convergence factor one obtains

$$V^t(\mathbf{p}' - \mathbf{p}) = -\frac{Z}{(2\pi)^{3/2}} \int d^3x \, \frac{j_0(|\mathbf{p}' - \mathbf{p}|r)}{r} = -\sqrt{\frac{2}{\pi}} \frac{Z}{|\mathbf{p}' - \mathbf{p}|^2} \tag{D.8}$$

Thus the momentum-space Schrödinger equation for a hydrogenlike atom becomes

$$(p'^2 + k^2)\chi^t(\mathbf{p}') = \frac{Z}{\pi^2} \int d^3p \, \frac{1}{|\mathbf{p}' - \mathbf{p}|^2} \chi^t(\mathbf{p}) \tag{D.9}$$

D.3 Projection of momentum-space onto a 4-dimensional hypersphere

In his famous 1935 paper, V.A. Fock, [Fock, 1935], [Fock, 1958], introduced the four unit vectors

$$u_j = \frac{2kp_j}{k^2 + p^2} \quad j = 1, 2, 3$$

$$u_4 = \frac{k^2 - p^2}{k^2 + p^2} \tag{D.10}$$

These unit vectors label the points on a 4-dimensional hypersphere, and the transformation defined by equation (D.10) projects 3-dimensional momentum-space onto the surface of the hypersphere. The momentum-space angles θ_p and ϕ_p are related to the Cartesian components of momentum, p_1, p_2 and p_3, by

$$p_1 = p \sin \theta_p \cos \phi_p$$

$$p_2 = p\sin\theta_p \sin\phi_p$$
$$p_3 = p\cos\theta_p \tag{D.11}$$

and the momentum-space volume element can be written in the form

$$d^3p = p^2 dp \, \sin\theta_p \, d\theta_p d\phi_p \tag{D.12}$$

In the 4-dimensional space in which the hypersphere is embedded, we can introduce another angle, χ_p, defined by

$$\frac{2kp}{k^2 + p^2} = \sin\chi_p$$
$$\frac{k^2 - p^2}{k^2 + p^2} = \cos\chi_p \tag{D.13}$$

and

$$u_1 = \sin\chi_p \sin\theta_p \cos\phi_p$$
$$u_2 = \sin\chi_p \sin\theta_p \sin\phi_p$$
$$u_3 = \sin\chi_p \cos\theta_p$$
$$u_4 = \cos\chi_p \tag{D.14}$$

Since

$$\frac{d}{dp}\cos\chi_p = -\sin\chi_p \frac{d\chi_p}{dp} \tag{D.15}$$

it follows from (D.13) that

$$\frac{d\chi_p}{dp} = \frac{2k}{k^2 + p^2} \tag{D.16}$$

Thus the element of solid angle $d\Omega$ in the 4-dimensional hyperspace is related to the momentum-space volume element by

$$d\Omega = \sin^2\chi_p \sin\theta_p \, d\chi d\theta_p d\phi_p = \left(\frac{2k}{k^2+p^2}\right)^3 d^3p \tag{D.17}$$

With a certain amount of work, one can also show that

$$\frac{1}{|\mathbf{p'} - \mathbf{p}|^2} = \frac{4k^2}{(k^2+p^2)(k^2+p'^2)}\frac{1}{|\mathbf{u'}-\mathbf{u}|^2} \tag{D.18}$$

and substituting (D.18) into (D.4) yields

$$(p'^2 + k^2)^2 \chi^t(\mathbf{p'}) = \frac{Z}{2k\pi^2}\int d\Omega \, \frac{(p^2+k^2)^2}{|\mathbf{u'}-\mathbf{u}|^2}\chi^t(\mathbf{p}) \tag{D.19}$$

If we next let

$$\chi^t(\mathbf{p}) = M(p)F(\mathbf{u}) \tag{D.20}$$

where

$$M(p) \equiv \frac{4k^{5/2}}{(k^2+p^2)^2} \tag{D.21}$$

the integral equation for hydrogenlike orbitals can be written in the simplified form

$$F(\mathbf{u'}) = \frac{Z}{2k\pi^2}\int d\Omega\, \frac{1}{|\mathbf{u'}-\mathbf{u}|^2}F(\mathbf{u}) \tag{D.22}$$

D.4 Expansion of the kernel in terms of hyperspherical harmonics

Fock next expanded the kernel of the integral equation (D.22) in terms of hyperspherical harmonics. From equation (10.32) with $d=4$ and

$$r_> = r_< = 1 \tag{D.23}$$

we have

$$\frac{1}{|\mathbf{u'}-\mathbf{u}|^2} = \sum_{\lambda=0}^{\infty} C_\lambda^1(\mathbf{u}\cdot\mathbf{u'}) \tag{D.24}$$

We can also make use of the sum rule (10.50) with $d=4$. This gives us

$$\frac{1}{|\mathbf{u'}-\mathbf{u}|^2} = \sum_{\lambda=0}^{\infty} \frac{2\pi^2}{\lambda+1} \sum_{l,m} Y_{\lambda,l,m}(\mathbf{u'})Y_{\lambda,l,m}^*(\mathbf{u}) \tag{D.25}$$

Substituting (D.25) into (D.22), we obtain

$$F(\mathbf{u'}) = \sum_{\lambda=0}^{\infty} \frac{Z}{k(\lambda+1)} \sum_{l,m} Y_{\lambda,l,m}(\mathbf{u'})\int d\Omega\, Y_{\lambda,l,m}^*(\mathbf{u})F(\mathbf{u}) \tag{D.26}$$

Then, using the orthonormality relation for the 4-dimensional hyperspherical harmonics,

$$\int d\Omega\, Y_{\lambda,l,m}^*(\mathbf{u})Y_{\lambda',l',m'}(\mathbf{u}) = \delta_{\lambda',\lambda}\delta_{l',l}\delta_{m',m} \tag{D.27}$$

we can see that the integral equation (D.26) will be solved if we let

$$F(\mathbf{u}) = Y_{\lambda',l',m'}(\mathbf{u}) \tag{D.28}$$

provided that
$$\frac{Z}{k(\lambda'+1)} = 1 \qquad (D.29)$$

Making the identification $\lambda' = n - 1$, and remembering that k is related to the energy by $k^2 = -2E$, we can see from (D.20), (D.28) and (D.29) that
$$\chi^t_{n,l,m}(\mathbf{p}) = M(p) Y_{n-1,l,m}(\mathbf{u}) \qquad (D.30)$$

will be a solution to the momentum-space wave equation for hydrogenlike atoms provided that
$$E = -\frac{k^2}{2} = -\frac{Z^2}{2n^2} \qquad n = 1, 2, 3, \ldots \qquad (D.31)$$

which is Fock's famous result.

Equation (D.26) can be rewritten in the even simpler form:
$$F(\mathbf{u}) = \sum_{\lambda=0}^{\infty} \frac{Z}{k(\lambda+1)} O_\lambda [F(\mathbf{u})] \qquad (D.32)$$

where O_λ is a projection operator that projects out a harmonic polynomial of order λ from $F(\mathbf{u})$. From (D.32) we can see that if
$$F(\mathbf{u}) = O_\lambda [F(\mathbf{u})] \qquad (D.33)$$
and
$$\frac{Z}{k(\lambda+1)} = 1 \qquad (D.34)$$

then $M(p)F(\mathbf{u})$ will correspond to a momentum-space solution to the hydrogenlike one-electron Schrödinger equation. Thus, any harmonic polynomial whatever in u_1, u_2, u_3 and u_4 corresponds to such a solution.

Appendix E

THE GREEN'S FUNCTION OF THE SCHRÖDINGER EQUATION

E.1 The operator $-\Delta + p_\kappa^2$ and its Green's function

If atomic units are used, the Schrödinger equation for a system of N electrons moving in the potential $V(\mathbf{x})$ can be written in the form

$$\left[-\Delta + p_\kappa^2 + 2V(\mathbf{x})\right] \Psi_\kappa(\mathbf{x}) = 0 \tag{E.1}$$

where

$$p_\kappa^2 \equiv -2E_\kappa \tag{E.2}$$

and

$$\Delta \equiv \sum_{j=1}^{d} \frac{\partial^2}{\partial x_j^2} \tag{E.3}$$

The operator Δ is the generalized Laplacian operator in a d-dimensional space, where $d = 3N$ and where N is the number of electrons in the system. In the notation of this appendix, $\mathbf{x}$ stands for the d-dimensional coordinate vector representing the positions of the N electrons, and

$$\Psi_\kappa(\mathbf{x}) \equiv \Psi_\kappa(x_1, \ldots, x_d)$$
$$dx \equiv dx_1 dx_2 \ldots dx_d$$
$$dp \equiv dp_1 dp_2 \ldots dp_d \tag{E.4}$$

We would like to show that the Green's function of $-\Delta + p_\kappa^2$ is given by

$$G(\mathbf{x} - \mathbf{x}') = \frac{1}{(2\pi)^d} \int dp \, \frac{e^{i\mathbf{p}\cdot(\mathbf{x}-\mathbf{x}')}}{p_\kappa^2 + p^2} \tag{E.5}$$

where $e^{i\mathbf{p}\cdot(\mathbf{x}-\mathbf{x}')}$ is a d-dimensional plane wave. Applying the operator $-\Delta + p_\kappa^2$ to both sides of equation (E.5), we obtain

$$\left[-\Delta + p_\kappa^2\right] G(\mathbf{x}-\mathbf{x}') = \frac{1}{(2\pi)^d} \int dp\, e^{i\mathbf{p}\cdot(\mathbf{x}-\mathbf{x}')} = \delta(\mathbf{x}-\mathbf{x}') \quad (\text{E.6})$$

Thus $G(\mathbf{x}-\mathbf{x}')$ is seen to be the Green's function of $-\Delta + p_\kappa^2$. To show that

$$\Psi_\kappa(\mathbf{x}) = -2 \int d\mathbf{x}'\, G(\mathbf{x}-\mathbf{x}') V(\mathbf{x}') \Psi_\kappa(\mathbf{x}') \quad (\text{E.7})$$

is an integral form of the Schrödinger equation (E.1), we act on both sides of (E.7) with $-\Delta + p_\kappa^2$. With the help of (E.6), this gives

$$\left[-\Delta + p_\kappa^2\right] \Psi_\kappa(\mathbf{x}) = -2\left[-\Delta + p_\kappa^2\right] \int d\mathbf{x}'\, G(\mathbf{x}-\mathbf{x}') V(\mathbf{x}') \Psi_\kappa(\mathbf{x}')$$

$$= -2 \int d\mathbf{x}'\, \delta(\mathbf{x}-\mathbf{x}') V(\mathbf{x}') \Psi_\kappa(\mathbf{x}')$$

$$= -2V(\mathbf{x}) \Psi_\kappa(\mathbf{x}) \quad (\text{E.8})$$

from which it follows that if $G(\mathbf{x}-\mathbf{x}')$ is defined by (E.5), then (E.7) is an integral form of the N-electron Schrödinger equation (E.1).

E.2 Conservation of symmetry under Fourier transformation

Suppose that $\mathcal{G}$ is a group of operations that leave the hyperradius invariant, and suppose that P_γ^ν is a group-theoretical projection operator corresponding to the νth basis function of the γth irreducible representation of $\mathcal{G}$. If we let $\tilde{P}_\gamma^\nu$ be the corresponding projection operator acting in momentum space, then

$$P_\gamma^\nu \left[(\mathbf{u}\cdot\mathbf{u}_p)^n\right] = \tilde{P}_\gamma^\nu \left[(\mathbf{u}_p\cdot\mathbf{u})^n\right] = \tilde{P}_\gamma^\nu \left[(\mathbf{u}\cdot\mathbf{u}_p)^n\right] \quad (\text{E.9})$$

The first equality in equation (E.9) holds because renaming the unit vectors has no effect, while the second equality holds because $\mathbf{u}\cdot\mathbf{u}_p = \mathbf{u}_p\cdot\mathbf{u}$. Combining equations (10.40) and (E.9) we can see that

$$\tilde{P}_\gamma^\nu \left[e^{-i\mathbf{p}\cdot\mathbf{x}}\right] = P_\gamma^\nu \left[e^{-i\mathbf{p}\cdot\mathbf{x}}\right] \quad (\text{E.10})$$

Now let us consider a function $f(\mathbf{x})$ which transforms under the elements of $\mathcal{G}$ like the νth basis function of the γth irreducible representation of $\mathcal{G}$.

Then
$$P_\gamma^\nu [f(\mathbf{x})] = f(\mathbf{x}) \tag{E.11}$$

Let us now calculate the d-dimensional Fourier transform of $f(\mathbf{x})$ making use of the idempotent property of projection operators:

$$\begin{aligned}
f^t(\mathbf{p}) &= \frac{1}{(2\pi)^{d/2}} \int dx\, e^{-i\mathbf{p}\cdot\mathbf{x}} f(\mathbf{x}) \\
&= \frac{1}{(2\pi)^{d/2}} \int dx\, e^{-i\mathbf{p}\cdot\mathbf{x}} P_\gamma^\nu [f(\mathbf{x})] \\
&= \frac{1}{(2\pi)^{d/2}} \int dx\, P_\gamma^\nu \left[e^{-i\mathbf{p}\cdot\mathbf{x}}\right] f(\mathbf{x}) \\
&= \frac{1}{(2\pi)^{d/2}} \int dx\, \tilde{P}_\gamma^\nu \left[e^{-i\mathbf{p}\cdot\mathbf{x}}\right] f(\mathbf{x}) \\
&= \tilde{P}_\gamma^\nu [f^t(\mathbf{p})]
\end{aligned} \tag{E.12}$$

Thus we see that symmetry with respect to a group of operations that leave the hyperradius invariant is conserved under a d-dimensional Fourier transformation.

E.3 Conservation of symmetry under Green's function iteration

Let us now consider a group of operations $\mathcal{G}$ that not only leave the hyperradius invariant but also commute with the potential $V(\mathbf{x})$. Iteration of the integral form of the N-particle Schrödinger equation (E.7) will then leave the symmetry of a trial function $\Psi_\kappa^i(\mathbf{x})$ invariant. The iteration involves a double Fourier transformation, but since symmetry is conserved under both these Fourier transformations, and since $(p_\kappa^2 + p^2)^{-1}$ is invariant, the symmetry of the trial function is conserved under Green's function iteration. To make this assertion more precise, we make the identification:

$$-V(\mathbf{x})\Psi_\kappa^i(\mathbf{x}) = f(\mathbf{x}) \tag{E.13}$$

If
$$P_\gamma^\nu \left[V(\mathbf{x})\Psi_\kappa^i(\mathbf{x})\right] = V(\mathbf{x})P_\gamma^\nu \left[\Psi_\kappa^i(\mathbf{x})\right] = V(\mathbf{x})\Psi_\kappa^i(\mathbf{x}) \tag{E.14}$$

then
$$P_\gamma^\nu [f(\mathbf{x})] = f(\mathbf{x}) \tag{E.15}$$

and

$$\tilde{P}^\nu_\gamma \left[f^t(\mathbf{p}) \right] = f^t(\mathbf{p}) \tag{E.16}$$

From equations (E.7) and (E.5) we have

$$\Psi^{i+1}_\kappa(\mathbf{x}) = 2 \int d x'\ G(\mathbf{x} - \mathbf{x}') f(\mathbf{x}')$$

$$= \frac{2}{(2\pi)^{d/2}} \int dp\ \frac{e^{i\mathbf{p}\cdot\mathbf{x}}}{p_\kappa^2 + p^2}\ f^t(\mathbf{p}) \tag{E.17}$$

Applying our projection operator to the iterated solution we have

$$P^\nu_\gamma \left[\Psi^{i+1}_\kappa(\mathbf{x}) \right] = \frac{2}{(2\pi)^{d/2}} \int dp\ \frac{P^\nu_\gamma \left[e^{i\mathbf{p}\cdot\mathbf{x}} \right]}{p_\kappa^2 + p^2}\ f^t(\mathbf{p})$$

$$= \frac{2}{(2\pi)^{d/2}} \int dp\ \frac{\tilde{P}^\nu_\gamma \left[e^{i\mathbf{p}\cdot\mathbf{x}} \right]}{p_\kappa^2 + p^2}\ f^t(\mathbf{p})$$

$$= \frac{2}{(2\pi)^{d/2}} \int dp\ \frac{e^{i\mathbf{p}\cdot\mathbf{x}}}{p_\kappa^2 + p^2}\ \tilde{P}^\nu_\gamma \left[f^t(\mathbf{p}) \right]$$

$$= \frac{2}{(2\pi)^{d/2}} \int dp\ \frac{e^{i\mathbf{p}\cdot\mathbf{x}}}{p_\kappa^2 + p^2}\ f^t(\mathbf{p})$$

$$= \Psi^{i+1}_\kappa(\mathbf{x}) \tag{E.18}$$

Thus we see that Green's function iteration preserves symmetry under $\mathcal{G}$.

E.4 Alternative representations of the Green's function

Because of the great importance of the Green's function of the operator $-\Delta + p_\kappa^2$, it is interesting to consider some alternative representations of this function. From equation (10.59) we have

$$\int d\Omega_p\ e^{i\mathbf{p}\cdot\mathbf{x}} = \int d\Omega\ e^{i\mathbf{p}\cdot\mathbf{x}} = (d-2)!!\, I(0) j_0^d(pr) \tag{E.19}$$

where $d\Omega_p$ is the generalized solid angle element in momentum space. The volume element in momentum space can be written in the form

$$dp = dp_1 dp_2 ... dp_d = dp\ p^{d-1} d\Omega_p \tag{E.20}$$

and thus

$$G(\mathbf{x}-\mathbf{x}') = \frac{1}{(2\pi)^d}\int dp\, \frac{e^{i\mathbf{p}\cdot(\mathbf{x}'-\mathbf{x})}}{p_\kappa^2+p^2}$$

$$= \frac{1}{(2\pi)^d}\int_0^\infty dp\, \frac{p^{d-1}}{p_\kappa^2+p^2}\int d\Omega_p\, e^{i\mathbf{p}\cdot(\mathbf{x}-\mathbf{x}')}$$

$$= \frac{(d-2)!!I(0)}{(2\pi)^d}\int_0^\infty dp\, \frac{p^{d-1}}{p_\kappa^2+p^2}\, j_0^d(p|\mathbf{x}-\mathbf{x}'|) \quad\text{(E.21)}$$

Equation (E.21) gives us an integral representation of the Green's function involving only a single integration, and it demonstrates that it is a function only of the distance $|\mathbf{x}-\mathbf{x}'|\equiv r$. If we let $pr = s$ and make use of equations (10.49) and (10.52), we can rewrite (E.21) in the form

$$G(\mathbf{x}-\mathbf{x}') = \frac{1}{(2\pi)^{\alpha+1}r^{2\alpha}}\int_0^\infty ds\, \frac{s^{\alpha+1}}{(p_\kappa r)^2+s^2}J_\alpha(s) \quad\text{(E.22)}$$

where α is defined by (10.44). The integral in equation (E.22) can be evaluated analytically, and the result is

$$G(\mathbf{x}-\mathbf{x}') = \frac{p_\kappa^\alpha K_\alpha(p_\kappa|\mathbf{x}-\mathbf{x}'|)}{2\pi(2\pi|\mathbf{x}-\mathbf{x}'|)^\alpha} \qquad \alpha \equiv \frac{d}{2}-1 \quad\text{(E.23)}$$

where K_α is a modified Bessel function of the second kind.

A second alternative representation of the Green's function can be obtained using the methods outlined in Appendix C, which yield an expansion in terms of Gegenbauer polynomials:

$$G(\mathbf{x}-\mathbf{x}') = \sum_{\lambda=0}^\infty g_\lambda(r,r')C_\lambda^\alpha(\mathbf{u}\cdot\mathbf{u}') \quad\text{(E.24)}$$

where

$$g_\lambda(r,r') = I(0)\alpha(\alpha+\lambda)\frac{(\Gamma(\alpha)2^\alpha)^2}{(2\pi)^{2\alpha+2}r^\alpha r'^\alpha}\int_0^\infty dp\, \frac{p}{p_\kappa^2+p^2}J_{\alpha+\lambda}(pr)J_{\alpha+\lambda}(pr') \quad\text{(E.25)}$$

If we make use of this expansion, and of the fact that for any function of $\mathbf{u}$,

$$\frac{\alpha+\lambda}{\alpha I(0)}\int d\Omega'\, C_\lambda^\alpha(\mathbf{u}\cdot\mathbf{u}')f(\mathbf{u}') = O_\lambda\left[f(\mathbf{u})\right] \quad\text{(E.26)}$$

we can rewrite the integral form of the N-electron Schrödinger equation (E.7) in the form

$$\Psi_\kappa(\mathbf{x}) = -2 \sum_{\lambda=0}^{\infty} \frac{\alpha I(0)}{\alpha + \lambda} \int_0^\infty dr'\, r'^{d-1} g_\lambda(r,r') O_\lambda \left[V(\mathbf{x}')\Psi_\kappa(\mathbf{x}')\right]_{u'=u} \quad (E.27)$$

For $N=1$ and $d=3$ this becomes

$$\Psi_\kappa(\mathbf{x}) = -2 \sum_{l=0}^{\infty} \frac{4\pi}{2l+1} \int_0^\infty dr'\, r'^{2} g_l(r,r') O_l \left[V(\mathbf{x}')\Psi_\kappa(\mathbf{x}')\right]_{u'=u} \quad (E.28)$$

where

$$g_l(r,r') = \frac{2l+1}{4\pi\sqrt{rr'}} \int_0^\infty dp\, \frac{p}{p_\kappa^2 + p^2} J_{l+\frac{1}{2}}(pr) J_{l+\frac{1}{2}}(pr') \quad (E.29)$$

or, expressed in terms of spherical Bessel functions,

$$g_l(r,r') = \frac{2l+1}{2\pi^2} \int_0^\infty dp\, \frac{p^2}{p_\kappa^2 + p^2} j_l(pr) j_l(pr') \quad (E.30)$$

The p integration in (E.29) or (E.30) can be evaluated by contour integration. Alternatively we can use a method discussed in Appendix C, equations (C.25)-(C.39), from which we have

$$g_l(r,r') = \frac{2l+1}{2rr'} \int_{|r-r'|}^{r+r'} dr''\, r'' G(r'') P_l\left(\frac{r^2 + r'^2 - r''^2}{2rr'}\right) \quad (E.31)$$

Substituting $\alpha = \frac{1}{2}$ into (E.23) yields

$$G(r'') = \frac{1}{2\pi} \sqrt{\frac{p_\kappa}{2\pi r''}} K_{\frac{1}{2}}(p_\kappa r'') = \frac{e^{-p_\kappa r''}}{4\pi r''} \quad (E.32)$$

so that (E.31) can be written in the form

$$g_l(r,r') = \frac{2l+1}{8\pi rr'} \int_{|r-r'|}^{r+r'} dr''\, e^{-p_\kappa r''} P_l\left(\frac{r^2 + r'^2 - r''^2}{2rr'}\right) \quad (E.33)$$

When $l=0$, (E.33) becomes

$$g_0(r,r') = \frac{1}{8\pi rr'} \int_{|r-r'|}^{r+r'} dr''\, e^{-p_\kappa r''}$$

$$= \frac{1}{8\pi p_\kappa rr'} \left(e^{-p_\kappa |r-r'|} - e^{-p_\kappa(r+r')}\right)$$

$$= \frac{p_\kappa}{4\pi} i_0(p_\kappa r_<) k_0(p_\kappa r_>) \quad (E.34)$$

where i_0 and k_0 are respectively modified spherical Bessel functions of the first and second kind, and where $r_<$ is the smaller of r and r', while $r_>$ is the greater. For general l we have

$$g_l(r, r') = (2l+1) P_l(\mathcal{W}) g_0(r, r') \tag{E.35}$$

where the operator $\mathcal{W}$ is defined by

$$\mathcal{W} \equiv \frac{r^2 + r'^2}{2rr'} - \frac{1}{2rr'} \frac{\partial^2}{\partial p_\kappa^2} \tag{E.36}$$

We can also write $g_l(r, r')$ in the form

$$g_l(r, r') = \frac{(2l+1)p_\kappa}{4\pi} i_l(p_\kappa r_<) k_l(p_\kappa r_>) \tag{E.37}$$

To understand why equation (E.37) holds, we act on the Green's function with the operator $-\Delta + p_\kappa^2$. When $d = 3$ and $|\mathbf{x} - \mathbf{x}'| \neq 0$ we have

$$\left[-\Delta + p_\kappa^2\right] G(\mathbf{x} - \mathbf{x}') = \sum_{l=0}^{\infty} \left[-\Delta + p_\kappa^2\right] g_l(r, r') P_l(\mathbf{u} \cdot \mathbf{u}') = 0 \tag{E.38}$$

Since each term in the sum must vanish separately, we have

$$\left[-\Delta + p_\kappa^2\right] g_l(r, r') P_l(\mathbf{u} \cdot \mathbf{u}') = 0 \tag{E.39}$$

from which it follows that

$$\left[r^2 \frac{\partial^2}{\partial r^2} + 2r \frac{\partial}{\partial r} - l(l+1) - (p_\kappa r)^2\right] g_l(r, r') = 0 \tag{E.40}$$

Similarly

$$\left[r'^2 \frac{\partial^2}{\partial r'^2} + 2r' \frac{\partial}{\partial r'} - l(l+1) - (p_\kappa r')^2\right] g_l(r, r') = 0 \tag{E.41}$$

Thus $g_l(r, r')$ obeys the modified spherical Bessel differential equation both with respect to r and with respect to r'. Equations (E.40) and (E.41) are therefore satisfied by the expression on the right-hand side of (E.37). The remaining details of (E.37) can be determined (for example) by (E.35) and (E.36).

Appendix F

CONFIGURATIONS BASED ON COULOMB STURMIAN ORBITALS

F.1 Coulomb Sturmian spin-orbitals

Most of the calculations discussed in this book have been based on isoenergetic generalized Sturmians configurations of the Goscinskian type. However, there exist many alternative possibilities for constructing isoenergetic configurations to be used in atomic calculations. In this appendix, we shall discuss one of these alternative methods - the use of configurations based on 1-electron Coulomb Sturmian orbitals of the type discussed in Chapter 1. In other words, we shall consider configurations of the form

$$\Phi_\nu(\mathbf{x}) = |\chi_{n,l,m,m_s} \chi_{n',l',m',m'_s} \chi_{n'',l'',m'',m''_s} \cdots| \qquad (F.1)$$

where the 1-electron spin-orbitals in the Slater determinant are defined by equations (1.11)-(1.21). The radial parts of the first few Coulomb Sturmian orbitals are shown in Table 1.1. The reader will remember from Chapter 1 that all of the orbitals in a Coulomb Sturmian basis set have the same exponential factor, e^{-kr}, the constant k being the same for the entire basis set. The Coulomb Sturmians have exactly the same form as hydrogenlike spin-orbitals, except that Z/n is everywhere replaced by k. Thus the Coulomb Sturmians obey the 1-electron wave equation

$$\left[-\frac{1}{2}\nabla_j^2 + \frac{1}{2}k^2 - \frac{nk}{r_j} \right] \chi_\mu(\mathbf{x}_j) = 0 \qquad (F.2)$$

where j is the electron index, and where we let μ stand for all four 1-electron indices, i.e.

$$\mu \equiv (n, l, m, m_s) \qquad (F.3)$$

It will also be remembered from Chapter 1 that the Coulomb Sturmians obey a potential-weighted orthonormality relation:

$$\int d\tau_j \, \chi_{\mu'}^*(\mathbf{x}_j)\frac{1}{r_j}\chi_\mu(\mathbf{x}_j) = \frac{k}{n}\delta_{\mu',\mu} \qquad (\text{F.4})$$

We would like to use the configurations defined in equations (F.1)-(F.4) to build up solutions to the non-relativistic Schrödinger equation for an atom or ion:

$$\left[-\frac{1}{2}\Delta + V(\mathbf{x}) - E_\kappa\right]\Psi_\kappa(\mathbf{x}) = 0 \qquad (\text{F.5})$$

where

$$\Delta \equiv \sum_{j=1}^N \nabla_j^2 \qquad (\text{F.6})$$

and where

$$V(\mathbf{x}) = -\sum_{j=0}^N \frac{Z}{r_j} + \sum_{i>j}^N\sum_{j=0}^N \frac{1}{r_{ij}} \qquad (\text{F.7})$$

Letting

$$\Psi_\kappa(\mathbf{x}) = \sum_\nu \Phi_\nu(\mathbf{x}) B_{\nu,\kappa} \qquad (\text{F.8})$$

and

$$E_\kappa = -\frac{Nk^2}{2} \qquad (\text{F.9})$$

we have

$$\sum_\nu \left[-\frac{1}{2}\Delta + \frac{Nk^2}{2} + V(\mathbf{x})\right]\Phi_\nu(\mathbf{x}) B_{\nu,\kappa} = 0 \qquad (\text{F.10})$$

At this point, the motivation behind equation (F.9) needs a little discussion: Because of the completeness properties of the Coulomb Sturmians, the results of the calculation are quite insensitive to the choice of k, provided that the basis set is sufficiently large. Since we are thus free to choose k as we like, we impose the condition (F.9) so that we can use the 1-electron equation (F.2) to simplify the secular equations.

F.2 Generalized Shibuya-Wulfman integrals

To see how this simplification takes place, we make the definition

$$\mathfrak{S}_{\nu',\nu} \equiv \int dx\ \Phi_{\nu'}^*(\mathbf{x}) \left(\frac{-\Delta + Nk^2}{2k^2} \right) \Phi_\nu(\mathbf{x}) \qquad (\text{F.11})$$

If we compare this definition with equation (11.10), we can see that the integrals $\mathfrak{S}_{\nu',\nu}$ can be thought of as N-electron, 1-center analogues of the Shibuya-Wulfman integrals. We shall see that the matrix elements $\mathfrak{S}_{\nu',\nu}$ defined by equation (F.11) are independent of k. Making use of (F.6) we can rewrite (F.11) as

$$\mathfrak{S}_{\nu',\nu} = \frac{1}{k^2} \int dx\ \Phi_{\nu'}^*(\mathbf{x}) \sum_{j=1}^N \left[-\frac{1}{2}\nabla_j^2 + \frac{k^2}{2} \right] \Phi_\nu(\mathbf{x}) \qquad (\text{F.12})$$

The next step is to notice that because the antisymmetrizer is idempotent, we can rewrite (F.12) in the form

$$\mathfrak{S}_{\nu',\nu} = \frac{\sqrt{N!}}{k^2} \int dx\ \Phi_{\nu'}^*(\mathbf{x}) \sum_{j=1}^N \left[-\frac{1}{2}\nabla_j^2 + \frac{k^2}{2} \right] \chi_{\mu_1}(\mathbf{x}_1)\chi_{\mu_2}(\mathbf{x}_2)\chi_{\mu_3}(\mathbf{x}_3)\cdots \qquad (\text{F.13})$$

Then from the 1-electron equation (F.2), it follows that

$$\mathfrak{S}_{\nu',\nu} = \frac{\sqrt{N!}}{k^2} \int dx\ \Phi_{\nu'}^*(\mathbf{x}) \left[\frac{n_1 k}{r_1} + \frac{n_2 k}{r_2} + \cdots \right] \chi_{\mu_1}(\mathbf{x}_1)\chi_{\mu_2}(\mathbf{x}_2)\chi_{\mu_3}(\mathbf{x}_3)\cdots \qquad (\text{F.14})$$

The potential-weighted orthonormality relations (F.4) can then be used to evaluate the integrals. For example, when $N=2$, we obtain

$$\begin{aligned}\mathfrak{S}_{\nu',\nu} =\ & \delta_{\mu_1',\mu_1}\langle\chi_{\mu_2'}|\chi_{\mu_2}\rangle + \delta_{\mu_2',\mu_2}\langle\chi_{\mu_1'}|\chi_{\mu_1}\rangle \\ & - \delta_{\mu_2',\mu_1}\langle\chi_{\mu_1'}|\chi_{\mu_2}\rangle - \delta_{\mu_1',\mu_2}\langle\chi_{\mu_2'}|\chi_{\mu_1}\rangle \end{aligned} \qquad (\text{F.15})$$

We now multiply equation (F.10) from the left by a conjugate configuration from our basis set and integrate over the coordinates. This gives us the secular equations

$$\sum_\nu \int dx\ \Phi_{\nu'}^*(\mathbf{x}) \left[-\frac{1}{2}\Delta + \frac{Nk^2}{2} + V(\mathbf{x}) \right] \Phi_\nu(\mathbf{x}) B_{\nu,\kappa} = 0 \qquad (\text{F.16})$$

If we let

$$T_{\nu',\nu} \equiv -\frac{1}{k} \int dx\ \Phi_{\nu'}^*(\mathbf{x}) V(\mathbf{x}) \Phi_\nu(\mathbf{x}) \qquad (\text{F.17})$$

then the secular equations take the form

$$\sum_{\nu}[T_{\nu',\nu} - k\mathfrak{S}_{\nu',\nu}] B_{\nu,\kappa} = 0 \qquad (F.18)$$

where, as usual, $T_{\nu',\nu}$ is energy-independent. The matrix elements of the nuclear attraction part of $T_{\nu',\nu}$ can be evaluated by means of the potential-weighted orthonormality relations (F.4). For example, when $N=2$, we obtain for the nuclear attraction term:

$$\begin{aligned}T^{(0)}_{\nu',\nu} =\ & \frac{Z}{n_1}\delta_{\mu'_1,\mu_1}\langle\chi_{\mu'_2}|\chi_{\mu_2}\rangle + \frac{Z}{n_2}\delta_{\mu'_2,\mu_2}\langle\chi_{\mu'_1}|\chi_{\mu_1}\rangle \\ & - \frac{Z}{n_1}\delta_{\mu'_2,\mu_1}\langle\chi_{\mu'_1}|\chi_{\mu_2}\rangle - \frac{Z}{n_2}\delta_{\mu'_1,\mu_2}\langle\chi_{\mu'_2}|\chi_{\mu_1}\rangle\end{aligned} \qquad (F.19)$$

F.3 An illustrative example

We can illustrate these results by considering the following very simple example: In the $N=2$ case, suppose that Z is so large that interelectron repulsion can be neglected. Then if we restrict our basis set to a single configuration with $\mu_1 \neq \mu_2$, we need only consider the diagonal matrix elements

$$\mathfrak{S}_{\nu,\nu} = 2 \qquad (F.20)$$

and

$$T^{(0)}_{\nu,\nu} = \frac{Z}{n_1} + \frac{Z}{n_2} \qquad (F.21)$$

The secular equation then requires that

$$\frac{Z}{n_1} + \frac{Z}{n_2} = 2k \qquad (F.22)$$

so that

$$4k^2 = \left[\frac{Z}{n_1} + \frac{Z}{n_2}\right]^2 \qquad (F.23)$$

This does not look like the result that we obtained using Goscinskian configurations and neglecting interelectron repulsion, except when $n_1 = n_2 = n$. In that special case it becomes

$$k^2 = \frac{Z^2}{n^2} = -E_\kappa \qquad (F.24)$$

Table F.1

	Gosc.	Coul.	Expt.
He	−2.89601	−2.90250	−2.90339
Li^+	−7.27139	−7.27836	−7.27984
Be^{2+}	−13.6467	−13.6538	−13.6565
B^{3+}	−22.0218	−22.0291	−22.0349
C^{4+}	−32.3970	−32.4043	−32.4158
N^{5+}	−44.7721	−44.7795	−44.8018
O^{6+}	−59.1471	−59.1546	−59.1922
F^{7+}	−75.5222	−75.5297	−75.5942
Ne^{8+}	−93.8972	−93.9048	−94.0055
Na^{9+}	−114.272	−114.280	−114.431
Mg^{10+}	−136.647	−136.655	−136.872
Al^{11+}	−161.022	−161.030	−161.334
Si^{12+}	−187.397	−187.405	−187.819
P^{13+}	−215.772	−215.780	−216.334
S^{14+}	−246.147	−246.155	−246.881
Cl^{15+}	−278.522	−278.530	−279.470

which agrees with our previous result. Thus we can see a difference between the Goscinskian configurations and configurations based on Coulomb Sturmians: With Goscinskian configurations, we can with reasonable accuracy represent a state by a single configuration, but with the configurations based on Coulomb Sturmians we can only do this in a few special cases. However, the basis set using Coulomb Sturmians has the compensating advantage that it is known to be complete.

F.4 Ground states for the 2-electron isoelectronic series

In practice one finds that the Coulomb Sturmian configurations are best for calculations of the ground states of few-electron atoms or ions, while the Goscinskian configurations are most useful for calculations of excited states. Some ground-state calculations for the 2-electron isoelectronic series are shown in Table F.1. A basis set consisting of 102 symmetry-adapted configurations was used for both calculations shown in the first column and those shown in the second. As can be seen from the table, the ground-state energies are better when Coulomb Sturmian configurations are used.

However, the Goscinskians give much better values for the excited states with this limited basis set.

F.5 Generalization to molecular problems

The concept of Coulomb Sturmian configurations can also be applied to molecular problems. To do this, we let the τ stand for five indices

$$\tau \equiv (n, l, m, m_s, a) \tag{F.25}$$

Here n, l, m and m_s have the same meanings as in equation (F.3), while a is the index of the particular nucleus on which a Coulomb Sturmian orbital is centered. Thus we can write

$$\chi_\tau(\mathbf{x}_j) \equiv \chi_\mu(\mathbf{x}_j - \mathbf{X}_a) \tag{F.26}$$

where μ is defined by (F.3). From (F.26) and (F.2), it follows that

$$\left[-\frac{1}{2}\nabla_j^2 + \frac{1}{2}k^2 - \frac{nk}{|\mathbf{x}_j - \mathbf{X}_a|} \right] \chi_\tau(\mathbf{x}_j) = 0 \tag{F.27}$$

We next define a molecular version of the Coulomb Sturmian configurations:

$$\Phi_\nu(\mathbf{x}) \equiv |\chi_\tau \chi_{\tau'} \chi_{\tau''} \cdots| \tag{F.28}$$

If we define the generalized Shibuya-Wulfman integrals by equation (F.11), remembering that $\Phi_\nu(\mathbf{x})$ is now given by (F.28), then equation (F.14) becomes

$$\mathfrak{S}_{\nu',\nu} = \frac{\sqrt{N!}}{k^2} \int d\mathbf{x}\ \Phi_{\nu'}^*(\mathbf{x}) \left[\frac{n_1 k}{|\mathbf{x}_1 - \mathbf{X}_{a_1}|} + \frac{n_2 k}{|\mathbf{x}_2 - \mathbf{X}_{a_2}|} + \cdots \right]$$
$$\times \chi_{\tau_1}(\mathbf{x}_1) \chi_{\tau_2}(\mathbf{x}_2) \chi_{\tau_3}(\mathbf{x}_3) \cdots \tag{F.29}$$

Just as in the case of atoms or atomic ions, the molecular Coulomb Sturmian secular equations have the form

$$\sum_\nu [T_{\nu',\nu} - k\mathfrak{S}_{\nu',\nu}] B_{\nu,\kappa} = 0 \tag{F.30}$$

where $T_{\nu',\nu}$ is defined by (F.17), and where the electronic energy levels of the molecule are related to k by

$$E_\kappa = -\frac{Nk^2}{2} \tag{F.31}$$

while the corresponding wave functions are

$$\Psi_\kappa(\mathbf{x}) = \sum_\nu \Phi_\nu(\mathbf{x}) B_{\nu,\kappa} \tag{F.32}$$

Notice that the molecular wave function is a linear combination of configurations built up from atomic orbitals, rather than a linear combination of configurations constructed from molecular orbitals.

Appendix G

NOTATION

$\nabla_j^2 \equiv$ the Laplacian operator of the jth electron

$N \equiv$ the number of electrons in a system

$\Delta \equiv \sum_{j=1}^{N} \nabla_j^2 \equiv$ the generalized Laplacian operator

$\Psi_\kappa \equiv$ a solution to the many electron wave equation

$E_\kappa \equiv$ the energy corresponding to Ψ_κ

$\Phi_\nu \equiv$ a generalized Sturmian basis function; usually a Goscinskian configuration

$\nu \equiv$ the set of quantum numbers characterizing Φ_ν

$B_{\nu\kappa} \equiv$ coefficients in the expansion $\Psi_\kappa = \sum_\nu \Phi_\nu B_{\nu\kappa}$

$\chi_\mu \equiv$ a one electron orbital

$\mu \equiv$ the set of quantum numbers characterizing χ_μ

$\epsilon_\mu \equiv$ the one electron energy corresponding to χ_μ

$V_0 \equiv$ the approximate potential used in constructing a set of generalized Sturmians

$V' \equiv$ the interelectron interaction potential

$V \equiv$ the actual potential

$\beta_\nu V_0 \equiv$ the weighted approximate potential

$\beta_\nu \equiv$ a set of weighting factors used to make the Sturmians isoenergetic

$Z \equiv$ the nuclear charge

$Q_\nu \equiv \beta_\nu Z \equiv$ the weighted nuclear charge characterizing the configuration Φ_ν

$p_\kappa \equiv \sqrt{-2E_\kappa} \equiv$ the scaling factor; a root of the Sturmian secular equation

$\mathcal{R}_\nu \equiv \sqrt{\dfrac{1}{n^2} + \dfrac{1}{n'^2} + \cdots}$ where $n, n', n'', \ldots$ appear in Φ_ν; $p_\kappa = Q_\nu \mathcal{R}_\nu$

$\mathcal{R}$-block $\equiv$ the set of configurations corresponding to a particular value of $\mathcal{R}_\nu$

$T^0_{\nu',\nu} \equiv -\dfrac{1}{p_\kappa} \int dx \Phi^*_{\nu'} V_0 \Phi_\nu = \mathcal{R}_\nu Z \delta_{\nu',\nu} \equiv$ the nuclear attraction matrix

$T'_{\nu',\nu} \equiv -\dfrac{1}{p_\kappa} \int dx \Phi^*_{\nu'} V' \Phi_\nu \equiv$ the energy independent interelectron repulsion matrix

$|\chi_\mu \chi_{\mu'} \cdots| \equiv$ a Slater determinant

$S \equiv$ a matrix of overlap integrals

$S_{ij} \equiv$ the matrix formed by deleting the ith row and jth column of S

$\mathcal{A} \equiv$ the antisymmetrizer

$P[\ldots] \equiv$ a projection operator acting on the quantity in square brackets

$P^\nu_\gamma[\ldots] \equiv$ a group theoretical projection operator

NOTATION

$\mathcal{E} \equiv$ an external electric field

$\mathcal{H} \equiv$ an external magnetic field

$(\sigma_x, \sigma_y, \sigma_z) \equiv$ the Pauli spin matrices

$(\alpha_x, \alpha_y, \alpha_z) \equiv$ the 4×4 Dirac alpha matrices

$(A_x, A_y, A_z, i\phi) \equiv A_\lambda \equiv$ the 4-component potential vector

$(j_x, j_y, j_z, ic\rho) \equiv j_\lambda \equiv$ the 4-component current vector

$\Omega_{j,l,M} \equiv$ a 2-component spherical spinor

$\tilde{\Omega}_{j,l,M} \equiv \Omega_{j,2j-l,M}$

$d \equiv$ the dimension of a space

$r \equiv \left(x_1^2 + x_2^2 + \ldots + x_d^2\right)^{1/2} \equiv$ the hyperradius

$d\Omega \equiv$ the element of solid angle in $\mathbb{R}^d$

$I(0) \equiv \int d\Omega \equiv$ total solid angle

$m_n \equiv$ a monomial of degree n

$f_n \equiv$ a homogeneous polynomial of degree n

$h_\lambda \equiv$ a harmonic polynomial of degree λ

$Y_{l,m} \equiv$ a spherical harmonic

$Y_{\lambda,\mu} \equiv$ a hyperspherical harmonic

$(\lambda, \mu) \equiv (\lambda, \mu_1, \mu_2, \ldots, \mu_{d-2})$ are the principal and minor quantum numbers of $Y_{\lambda,\mu}$

$\Lambda^2 \equiv$ the generalized angular momentum operator;
$\Lambda^2 Y_{\lambda,\mu} = \lambda(\lambda + d - 2)Y_{\lambda,\mu}$

$O_\lambda[\ldots] \equiv$ an operator that projects out eigenfunctions of Λ^2

$C_\lambda^\alpha \equiv$ a Gegenbauer polynomial

$$\alpha \equiv \frac{d}{2} - 1$$

$j_\lambda^d(t) \equiv \dfrac{\Gamma(\alpha)2^{\alpha-1} J_{\lambda+\alpha}(t)}{(d-4)!! t^\alpha}$ is a hyperspherical Bessel function

$k \equiv$ the common exponent of a set of Coulomb Sturmians

$p \equiv$ the momentum of an electron

$M(p) \equiv \dfrac{4k^{5/2}}{(k^2+p^2)^2}$ is Fock's universal factor

$\mathbf{X}_a \equiv$ the position of a nucleus

$\chi_\tau(\mathbf{x}) \equiv \chi_\mu(\mathbf{x}-\mathbf{X}_a)$ is a Coulomb Sturmian centered on $\mathbf{X}_a$

$\tau \equiv (n,l,m,m_s,a)$

$\mathfrak{S}_{\tau',\tau} \equiv \displaystyle\int d^3x\, \chi_{\tau'}^*(\mathbf{x}) \left(\dfrac{-\nabla^2+k^2}{2k^2} \right) \chi_\tau(\mathbf{x})$ is a Shibuya Wulfman integral

$K_{\tau',\tau} \equiv \sqrt{\dfrac{Z_{a'} Z_a}{n'n}} \mathfrak{S}_{\tau',\tau}$

$v(\mathbf{x}) \equiv \displaystyle\sum_a \dfrac{Z_a}{|\mathbf{x}-\mathbf{X}_a|}$ is a many center potential

$\mathfrak{W}_{\tau',\tau} \equiv -\dfrac{1}{k} \displaystyle\int d^3x\, \chi_{\tau'}^*(\mathbf{x}) v(\mathbf{x}) \chi_\tau(\mathbf{x})$ is a Wulfman integral

$\mathfrak{S}_{\nu',\nu} \equiv \displaystyle\int dx\, \Phi_{\nu'}^*(\mathbf{x}) \left(\dfrac{-\Delta + Nk^2}{2k^2} \right) \Phi_\nu(\mathbf{x})$ (where $\Phi_\nu = |\chi_\tau \chi_{\tau'} \cdots|$) is a generalized Shibuya Wulfman integral

Bibliography

Ahlberg, R. and Lindner, P., *The Fermi correlation for electrons in momentum space*, J. Phys. B, **9** (17) 2963-9, 1976.
Akhiezer, A.I. and Berestetskii, V.B., **Quantum Electrodynamics**, Interscience, New York, 1965.
Alliluev, S.P., Sov. Phys. JETP, **6** 156, 1958.
Amiet, J.-P. and Huguenin, P., **Mécaniques classique et quantiques dans l'espace de phase**, Université de Neuchâtel, 1981.
Amos, A.T. and Hall, G.G., Proc. Roy. Soc. London, **A263**, 483, 1961.
Anderson, R.W., Aquilanti, V., Cavalli, S. and Grossi, G., J. Phys. Chem., **95** 8184, 1991.
Anderson, R.W., Aquilanti, V., Cavalli, S. and Grossi, G., J. Phys. Chem., **97** 2443, 1993.
Aquilanti, V. and Cavalli, S., *Coordinates for molecular dynamics*, J. Chem. Phys., **85** 1355-1361, 1986.
Aquilanti, V., Cavalli, S., De Fazio, D. and Grossi, G. *Hyperangular momentum: Applications to atomic and molecular science*, in **New Methods in Quantum Theory**, Tsipis, C.A., Popov, V.S., Herschbach, D.R., and Avery, J.S., Eds., Kluwer, Dordrecht, 1996.
Aquilanti, V., Cavalli, S. and Grossi, G., *Hyperspherical coordinates for molecular dynamics by the method of trees and the mapping of potential-energy surfaces for triatomic systems*, J. Chem. Phys., **85** 1362, 1986.
Aquilanti, V., Grossi, G., Laganá, A., Pelikan, E. and Klar, H., *A decoupling scheme for a 3-body problem treated by expansions into hyperspherical harmonics. The hydrogen molecular ion*, Lett. Nuovo Cimento, **41**, 541, 1984.
Aquilanti, V., Grossi, G. and Laganá, A., *On hyperspherical mapping and harmonic expansions for potential energy surfaces*, J. Chem. Phys., **76** 1587-8, 1982.
Aquilanti, V., Laganá, A. and Levine, R.D., Chem. Phys. Lett., **158** 87, 1989.
Aquilanti, V. and Cavalli, S., Chem. Phys. Lett., **141** 309, 1987.
Aquilanti, V., Cavalli, S., Grossi, G., Rosi, M., Pellizzari, V., Sgamellotti, A. and Tarantelli, F., Chem. Phys. Lett., **16** 179, 1989.
Aquilanti, V., Cavalli, S., Grossi, G. and Anderson, R.W., J. Chem. Soc. Faraday

Trans., **86** 1681, 1990.

Aquilanti, V., Benevente, L., Grossi, G. and Vecchiocattivi, F., *Coupling schemes for atom-diatom interactions, and an adiabatic decoupling treatment of rotational temperature effects on glory scattering*, J. Chem. Phys., **89** 751-761, 1988.

Aquilanti, V. and Grossi, G., *Angular momentum coupling schemes in the quantum mechanical treatment of P-state atom collisions* J. Chem. Phys., **73** 1165-1172, 1980.

Aquilanti, V., Cavalli, S. and Grossi, G., Theor. Chem. Acta, **79** 283, 1991.

Aquilanti, V. and Cavalli, S., Few Body Systems, Suppl. **6** 573, 1992.

Aquilanti, V. and Grossi, G., Lett. Nuovo Cimento, **42** 157, 1985.

Aquilanti, V. Cavalli, S. and De Fazio, D., *Angular and hyperangular momentum coupling coefficients as Hahn polynomials*, J. Phys. Chem. **99** 15694, 1995.

Aquilanti, V., Cavalli, S., Coletti, C. and Grossi, G., *Alternative Sturmian bases and momentum space orbitals; an application to the hydrogen molecular ion*, Chem. Phys. **209** 405, 1996.

Aquilanti, V., Cavalli, S. and Coletti, C., *The d-dimensional hydrogen atom; hyperspherical harmonics as momentum space orbitals and alternative Sturmian basis sets*, Chem. Phys. **214** 1, 1997.

Aquilanti, V. and Avery, J., *Generalized potential harmonics and contracted sturmians*, Chem. Phys. Letters, **267** 1, 1997.

Aquilanti, V., Cavalli, S., Coletti, C., Di Domenico, D. and Grossi, G., Int. Rev. Phys. Chem., **20** 673, 2001.

Aquilanti, V. and Caligiana, A., Chem. Phys. Letters, **366** 157, 2002.

Aquilanti, V., Caligiana A. and Cavalli, S., Int. J. Quantum Chem., **92** 99, 2003.

Aquilanti, V., Caligiana, A., Cavalli, S. and Coletti, C., Int. J. Quantum Chem., **92** 212, 2003.

Aquilanti, V. and Caligiana, A., in **Fundamental World of Quantum Chemistry: A Tribute to the Memory of P.O. Löwdin, I**, E.J. Brändas and E.S. Kryachko, Eds., Kluwer, Dordrecht, 297, 2003.

Aquilanti, V. and Avery, J., *Sturmian expansions for quantum mechanical many-body problems and hyperspherical harmonics*, Adv. Quant. Chem., **39** 72-101, 2001.

Avery, J., **Creation and Annihilation Operators**, McGraw-Hill, 1976.

Avery, J. and Ørmen, Per-Johan, Int. J. Quantum Chem. **18** 953, 1980.

Avery, J., **Hyperspherical Harmonics; Applications in Quantum Theory**, Kluwer Academic Publishers, Dordrecht, 1989.

Avery, J., *Hyperspherical Sturmian basis functions in reciprocal space*, in **New Methods in Quantum Theory**, Tsipis, C.A., Popov, V.S., Herschbach, D.R. and Avery, J.S., Eds., Kluwer, Dordrecht, 1996.

Avery, J. and Antonsen, F., *A new approach to the quantum mechanics of atoms and small molecules*, Int. J. Quantum Chem., Symp. **23** 159, 1989.

Avery, J. and Antonsen, F., *Iteration of the Schrödinger equation, starting with Hartree-Fock wave functions*, Int. J. Quantum Chem., **42** 87, 1992.

Avery, J. and Antonsen, F., Theor. Chim. Acta, **85** 33, 1993.

Avery, J. and Herschbach, D. R., *Hyperspherical Sturmian basis functions*, Int. J.

Quantum Chem., **41** 673, 1992.

Avery, J. and Wen, Z.-Y., *A Formulation of the quantum mechanical many-body in terms of hyperspherical coordinates*, Int. J. Quantum Chem. **25** 1069, 1984.

Avery, J., *Correlation in iterated solutions of the momentum-space Schrödinger equation*, Chem. Phys. Lett., **138** (6) 520-4, 1987.

Avery, J., *Hyperspherical harmonics; Some properties and applications*, in **Conceptual Trends in Quantum Chemistry**, Kryachko, E.S., and Calais, J.L., Eds, Kluwer, Dordrecht, 1994.

Avery, J., Hansen, T.B., Wang, M. and Antonsen, F., *Sturmian basis sets in momentum space*, Int. J. Quantum Chem. **57** 401, 1996.

Avery, J. and Hansen, T.B., *A momentum-space picture of the chemical bond* Int. J. Quantum Chem. **60** 201, 1996.

Avery, J., *Many-particle Sturmians*, J. Math. Chem., **21** 285, 1997.

Avery, J. and Antonsen, F., *Relativistic sturmian basis functions*, J. Math. Chem. **24** 175, 1998.

Avery, J., *A formula for angular and hyperangular integration*, J. Math. Chem., **24** 169, 1998.

Avery, J., *Many-electron Sturmians applied to atoms and ions*, J. Mol. Struct. **458** 1, 1999.

Avery, J., *Many-electron Sturmians as an alternative to the SCF-CI Method*, Adv. Quantum Chem., **31** 201, 1999.

Avery, J., **Hyperspherical Harmonics and Generalized Sturmians**, Kluwer Academic Publishers, Dordrecht, Netherlands, 196 pages, 2000.

Avery, J. and Antonsen, F., *Magnetic interactions and 4-currents*, J. Molecular Structure (Theochem), **261**, 69, 1992.

Avery, J., Antonsen, F. and Shim, I., *4-Currents in relativistic quantum chemistry*, Int. J. Quantum Chem. **45**, 573, 1993.

Avery, J. *Selected applications of hyperspherical harmonics in quantum theory*, J. Phys. Chem. **97**, 2406, 1993.

Avery, J. and Antonsen, F., *Evaluation of angular integrals by harmonic projection*, Theor. Chim. Acta. **85**, 33, 1993.

Avery, J., *Fock transforms in reciprocal-space quantum theory*, J. Math. Chem. **15** 233, 1994.

Aquilanti V. and Avery, J., *Generalized potential harmonics and contracted Sturmians*, Chem. Phys. Lett. **267** 1-8, 1997.

Avery, J., *Many-electron Sturmians applied to atoms and ions*, J. Mol. Struct. (Theochem) **458** 1-9, 1999.

Avery, J., *A formula for angular and hyperangular integration*, J. Math. Chem., **24** 169-174, 1998.

Avery, J. and Sauer, S., *Many-electron Sturmians applied to molecules*, in **Quantum Systems in Chemistry and Physics, Volume 1**, A. Hernández-Laguna, J. Maruani, R. McWeeney and S. Wilson editors, Kluwer Academic Publishers, 2000.

Avery, J. and Coletti, C., *Generalized Sturmians applied to atoms in strong external fields*, J. Math. Chem. **27** 43-51, 2000.

Avery, J. and Coletti, C., *Many-electron Sturmians applied to atoms and ions in strong external fields*, in **New Trends in Quantum Systems in Chemistry and Physics**, J. Marauani et al. (eds.), pages 77-93, Kluwer Academic Publishers, 2001.

Avery, J. and Shim, R., *Core ionization energies of atoms and molecules calculated using the Generalized Sturmian Method*, Int. J. Quantum Chem. **79** 1-7, 2000.

Avery, J., *The Generalized Sturmian Method and inelastic scattering of fast electrons*, J. Math. Chem. **27**, 279-292, 2000.

Avery, J., *Sturmian methods in quantum theory*, Proc. Workshop on Concepts in Chemical Physics, G.D. Billing and N. Henriksen, editors, Danish Technical University, 2001.

Avery, J. and Shim, R., *Molecular Sturmians, Part 1*, Int. J. Quantum Chem., **83**, 1-10, 2001.

Avery, J., *Sturmians*, in **Handbook of Molecular Physics and Quantum Chemistry**, S. Wilson, ed., Wiley, Chichester, 2003.

Avery, J., *Harmonic polynomials, hyperspherical harmonics, and Sturmians*, in **Fundamental World of Quantum Chemistry, A Tribute Volume to the Memory of Per-Olov Löwdin**, E.J. Brändas and E.S. Kryachko, eds., Kluwer Academic Publishers, Dordrecht, 261-296, 2003.

Avery, J. and Avery, J., *The Generalized Sturmian Method for calculating spectra of atoms and ions*, J. Math. Chem., **33** 145-162, 2003.

Avery, J. and Avery, J., *Kramers pairs in configuration interaction*, Adv. Quant. Chem., **43** 185-206, 2003.

Avery, J., *Many-center Coulomb Sturmians and Shibuya-Wulfman integrals*, Int. J. Quantum Chem., 2003.

Avery, J., Avery J. and Goscinski, O., *Natural orbitals from generalized Sturmian calculations*, Adv. Quant. Chem., **43** 207-16, 2003.

Avery, J., Avery, J., Aquilanti, V. and Caligiana, A., *Atomic densities, polarizabilities and natural orbitals derived from generalized Sturmian calculations*, Adv. Quant. Chem., **47** 156-173, 2004.

Avery, J. and Avery, J., *Generalized Sturmian solutions for many-particle Schrödinger equations*, J. Phys. Chem. A, **41** 8848, 2004.

Avery, J. and Avery, J., *The Generalized Sturmian Library*, http://sturmian.kvante.org, 2006.

Ballot, L. and Farbre de la Ripelle, M., *Application of the hyperspherical formalism to trinucleon bound-state problems*, Ann. Phys., **127** 62, 1980.

Bandar, M. and Itzyksen, C., *Group theory and the H atom*, Rev. Mod. Phys. **38** 330, 346, 1966.

Bang, J.M. and Vaagen, J.S., *The Sturmian expansion: a well-depth-method for orbitals in a deformed potential*, Z. Phys. A, **297** (3) 223-36, 1980.

Bang, J.M., Gareev, F.G., Pinkston, W.T. and Vaagen, J.S., Phys. Rep. **125** 253-399, 1985.

Baryudin, L. E. and Telnov, D. A., *Sturmian expansion of the electron density deformation for 3d-metal ions in electric field*, Vestn. Leningr. Univ., Ser 4: Fiz., Khim. (1), p 83-6, 1991.

Baretty, Reinaldo; Ishikawa, Yasuyuki; and Nieves, Jose F., *Momentum space approach to relativistic atomic structure calculations*, Int. J. Quantum Chem., Quantum Chem. Symp., **20** 109-17, 1986.

Benesch, Robert and Smith, Vedene H. Jr., *Natural orbitals in momentum space and correlated radial momentum distributions. I. The 1S ground state of $Li+$*, Int. J. Quantum Chem., Symp, **4** 131-8, 1971.

Berry, R.S., Ceraulo, S.C. and Batka, J., *Valence electrons in atoms: Collective or independent-particle-like?*, in Herschbach, D.R., Avery, J. and Goscinski, O., Eds., **Dimensional Scaling in Chemical Physics**, Kluwer, Dordrecht, 1993.

Bethe, H.A. and Salpeter, E.E., **Quantum Mechanics of One- and Two-Electron Atoms**, equation (43.8), Plenum, 1977.

Biedenharn, L.C. and Louck, J.D., **Angular Momentum in Quantum Physics**, Addison Wesley, Reading, Mass, 1981.

Biedenharn, L.C. and Louck, J.D., **The Racah-Wigner Algebra in Quantum Theory**, Addison Wesley, Reading, Mass, 1981.

Blinder, S.M., *On Green's functions, propagators, and Sturmians for the nonrelativistic Coulomb problem*, Int. J. Quantum Chem., Quantum Chem. Symp., **18** 293-307, 1984.

Bransden, B.H., Noble, C.J. and Hewitt, R.N., *On the reduction of momentum space scattering equations to Fredholm form*, J. Phys. B: At., Mol., Opt. Phys., **26** (16) 2487-99, 1993.

Brion, C.E., *Looking at orbitals in the laboratory: the experimental investigation of molecular wave functions and binding energies by electron momentum spectroscopy*, Int. J. Quantum Chem., **29** (5) 1397-428, 1986.

Brink, D.M. and Satchler, G.R., **Angular Momentum**, Oxford University Press, 1968.

Calais, J-L., Defranceschi, M., Fripiat, J.G. and Delhalle, J., *Momentum space functions for polymers*, J. Phys.: Condens. Matter, **4** (26) 5675-91, 1992.

Calais, J.-L., *Fukutome classes in momentum space*, Theor. Chim. Acta, **86** (1-2) 137-47, 1993.

Calais, J.-L., *Orthogonalization in momentum space*, Int. J. Quantum Chem., **35** (6) 735-43, 1989.

Calais, J.-L., *Pathology of the Hartree-Fock method in configuration and momentum space*, J. Chim. Phys. Phys.-Chim. Biol., **84** (5) 601-6, 1987.

Caligiana, Andreia, **Sturmian Orbitals in Quantum Chemistry**, Ph.D. thesis, University of Perugia, Italy, October, 2003.

Chen, J. Cheng Y. and Ishihara, T., *Hydrogenic- and Sturmian-function expansions in three-body atomic problems*, Phys. Rev., **186** (1) 25-38, 1969.

Chiu, T.W., *Non-relativistic bound-state problems in momentum space*, J. Phys. A: Math. Gen. **19** (13) 2537-47, 1986.

Cinal, M., *Energy functionals in momentum space: exchange energy, quantum corrections, and the Kohn-Sham scheme*, Phys. Rev. A, **48** (3) 1893-902, 1993.

Clark, C.W. and Taylor, K.T., *The quadratic Zeeman effect in hydrogen Rydberg series: application of Sturmian functions*, J. Phys. B, **15** (8) 1175-93, 1982.

Clementi, E., J. Chem. Phys. **38** 996, 1963.
Cohen, L., *Generalized phase-space distribution functions*, J. Math. Phys., **7** 781-786, 1966.
Cohen, L. and Lee, C., *Correlation hole and physical properties: a model calculation*, Int. J. Quantum Chem., **29** (3) 407-24, 1986.
Coleman, A.J. and Yukalov, V.I., **Reduced Density Matrices: Coulson's Challenge**, Springer-Verlag, New York, 2000.
Coletti, C., **Struttura Atomica e Moleculare Come Rottura della Simmetria Ipersferica**, Ph.D. thesis, Chemistry Department, University of Perugia, Italy, 1998.
Collins, L.A. and Merts, A.L., *Atoms in strong, oscillating electric fields: momentum-space solutions of the time-dependent, three-dimensional Schrödinger equation*, J. Opt. Soc. Am. B: Opt. Phys., **7** (4) 647-58, 1990.
Coulson, C.A., *Momentum distribution in molecular systems. I. Single bond. III. Bonds of higher order*, Proc. Camb. Phys. Soc., **37** 55, 74, 1941.
Coulson, C.A. and Duncanson, W.E., *Momentum distribution in molecular systems. II. C and C-H bond*, Proc. Camb. Phys. Soc., **37** 67, 1941.
Dahl, J.P., *The Wigner function*, Physica A, **114** 439, 1982.
Dahl, J.P., *On the group of translations and inversions of phase space and the Wigner function*, Phys. Scripta, **25** 499-503, 1982.
Dahl, J.P., *Dynamical equations for the Wigner functions*, in **Energy Storage and Redistribution in Molecules**, p 557-571, Ed. J. Hinze, Plenum, New York, 1983.
Dahl, J.P., *The phase-space representation of quantum mechanics and the Bohr-Heisenberg correspondence principle*, in **Semiclassical Description of Atomic and Nuclear Collisions**, p 379-394, Eds. Bang, J. and De Boer, J., North Holland, Amsterdam, 1985.
Dahl, J.P., *The dual nature of phase-space representations*, in **Classical and Quantum Systems**, p 420-423, Eds: Doebner, H.D. and Schroeck, F., Jr., World Scientific, Singapore, 1993.
Dahl, J.P., *A phase space essay*, in **Conceptual Trends in Quantum Chemistry**, p 199-224, Eds: Kryachko, E.S. and Calais, J.L., Kluwer Academic Publishers, Dordrecht, Netherlands, 1994.
Dahl, J.P. and Springborg, M., *The Morse oscillator in position space, momentum space, and phase space* J. Chem. Phys., **88** (7) 4535-47, 1988.
Dahl, J.P. and Springborg, M., *Wigner's phase-space function and atomic structure. I. The hydrogen atom*, J. Mol. Phys., **47** 1001, 1982.
Das, G.P., Ghosh, S.K. and Sahni, V.C., *On the correlation energy density functional in momentum space*, Solid State Commun, **65** (7) 719-21, 1988.
D.son, E.R., **Reduced Density Matrices**, Academic Press, New York, 1976.
Davies, R.W. and Davies, K.T.R., *On the Wigner distribution function for an oscillator*, Ann. Phys., **89** 261-273, 1975.
De Prunele, E. *O(4,2) coherent states and hydrogenic atoms*, Phys. Rev. A, **42** (5) 2542-9, 1990.
De Windt, L., Defranceschi, M. and Delhalle, J., *Variation-iteration method in momentum space: determination of Hartree-Fock atomic orbitals*, Int. J.

Quantum Chem., **45** (6) 609-18, 1993.

Defranceschi, M.; Suard, M. and Berthier, G., *Numerical solution of Hartree-Fock equations for a polyatomic molecule: linear triatomic hydrogen in momentum space*, Int. J. Quantum Chem., **25** (5) 863-7, 1984.

Defranceschi, M.; Suard, M.; and Berthier, G., *Epitome of theoretical chemistry in momentum space*, Folia Chim. Theor. Lat., **18** (2) 65-82, 1990.

Defranceschi, M., *Theoretical investigations of the momentum densities for molecular hydrogen*, Chem. Phys., **115** (3) 349-58, 1987.

Defranceschi, M. and Delhalle, J., *Numerical solution of the Hartree-Fock equations for quasi-one- dimensional systems: prototypical calculations on the (hydrogen atom) x chain*, Phys. Rev. B: Condens. Matter, **34** (8, Pt. 2) 5862-73, 1986.

Defranceschi, M. and Delhalle, J., *Momentum space calculations on the helium atom*, Eur. J. Phys., **11** (3) 172-8, 1990.

Delande, D. and Gay, J.C., *The hydrogen atom in a magnetic field. Spectrum from the Coulomb dynamical group approach*, J. Phys. B: At. Mol. Phys., **19** (6) L173-L178, 1986.

Delhalle, J. and Defranceschi, M., *Toward fully numerical evaluation of momentum space Hartree-Fock wave functions. Numerical experiments on the helium atom*, Int. J. Quantum Chem., Quantum Chem. Symp., **21** 425-33, 1987.

Delhalle, J.; Fripiat, J.G.; and Defranceschi, M., *Improving the one-electron states of ab initio GTO calculations in momentum space. Tests on two-electron systems: hydride, helium, and lithium(1+)*, Bull. Soc. Chim. Belg., **99** (3) 135-45, 1990.

Delhalle, J. and Harris, F.E., *Fourier-representation method for electronic structure of chainlike systems: restricted Hartree-Fock equations and applications to the atomic hydrogen (H)x chain in a basis of Gaussian functions*, Phys. Rev. B: Condens. Matter, **31** (10) 6755-65, 1985.

Deloff, A. and Law, J., *Sturmian expansion method for bound state problems*, Phys. Rev. C, **21** (5) 2048-53, 1980.

Denteneer, P.J.H. and Van Haeringen, W., *The pseudopotential-density-functional method in momentum space: details and test cases*, J. Phys. C, **18** (21) 4127-42, 1985.

Desclaux, J.P., Comput. Phys. Commun., **9** 31, 1975.

Dirac, P.A.M., *Note on exchange phenomena in the Thomas atom*, Proc. Camb. Phil. Soc., **26** 376-385, 1930.

Dorr, M.; Potvliege, R.M.; and Shakeshaft, R., *Atomic hydrogen irradiated by a strong laser field: Sturmian basis calculations of rates for high-order multiphoton ionization, Raman scattering, and harmonic generation*, J. Opt. Soc. Am. B: Opt. Phys., **7** (4) 433-48, 1990.

Douglas, M.., *Coulomb perturbation calculations in momentum space and application to quantum-electrodynamic hyperfine-structure corrections*, Phys. Rev. A, **11** (5) 1527-38, 1975.

Drake, G.W.F. and Goldman, S.P., *Relativistic Sturmian and finite basis set methods in atomic physics*, Adv. At. Mol. Phys., **25** 393-416, 1988.

Dube, L.J. and Broad, J.T., *Sturmian discretization. II. The off-shelf Coulomb wavefunction*, J. Phys. B: At., Mol., Opt. Phys., **23** (11) 1711-32, 1990.

Dube, Louis J. and Broad, J. T., *Sturmian discretization: the off-shell Coulomb wave function*, J. Phys. B: At., Mol. Opt. Phys., **22** (18) L503, 1989.

Duchon, C; Dumont-Lepage, M.C.; and Gazeau, J.P., *On two Sturmian alternatives to the LCAO method for a many-center one-electron system*, J. Chem. Phys., **76** (1) 445-7, 1982.

Duchon, C.; Dumont-Lepage, M.C.; and Gazeau, J.P., *Sturmian methods for the many-fixed-centers Coulomb potential*, J. Phys. A: Math. Gen., **15** (4) 1227-41, 1982.

Duffy, P.; Casida, M.E; Brion, C.E; and Chong, D.P., *Assessment of Gaussian-weighted angular resolution functions in the comparison of quantum-mechanically calculated electron momentum distributions with experiment* Chem. Phys., **159** (3) 347-63, 1992.

Duncanson, W.E., *Momentum distribution in molecular systems. IV. H molecule ion, H_2^+*, Proc. Camb. Phil. Soc., **37** 47, 1941.

Dunlap, B.I., Chem. Phys. Lett., **30** 39, 1975.

Edmonds, A.R., **Angular Momentum in Quantum Chemistry**, Princeton University Press, 1960.

Edmonds, A.R., *Quadratic Zeeman effect. I. Application of the sturmian functions*, J. Phys. B, **6** (8) 1603-15, 1973.

Englefield, M.J., **Group theory and the Coulomb problem**, Wiley-Interscience, New York, 1972.

Epstein, P.S., Proc. Natl. Acad. Sci. (USA), **12** 637, 1926.

Eyre, D.and Miller, H.G., *Sturmian projection and an L2 discretization of three-body continuum effects*, Phys. Rev. C: Nucl. Phys., **32** (3) 727-37, 1985.

Eyre, D. and Miller, H.G., *Sturmian approximation of three-body continuum effects*, Phys. Lett. B, **153B** (1-2) 5-7, 1985.

Eyre, D. and Miller, H.G., *Sturmian expansion approximation to three-body scattering*, Phys. Lett. B, **129B** (1-2) 15-17, 1983.

Fano, U., *Wave propagation and diffraction on a potential ridge*, Phys. Rev. **A22** 2660, 1980.

Fano, U., *Unified treatment of collisions*, Phys. Rev. **A24** 2402, 1981.

Fano, U., *Correlations of two excited electrons*, Rep. Prog. Phys. **46** 97, 1983.

Fano, U. and Rao, A.R.P., **Atomic Collisions and Spectra**, Academic Press, Orlando, Florida, 1986.

Fernández Rico, J., Ramírez, G., López, R., and Fernández Alonso, J.I., Collect. Czech. Chem. Comm., **53** 2250, 1987.

Fernández Rico, J., López, R., Ema, I., and Ramírez, G., preprints, 1997.

Flores, J.C., *Kicked quantum rotator with dynamic disorder: a diffusive behavior in momentum space*, Phys. Rev. A, **44** (6) 3492-5, 1991.

Fock, V.A., Z. Phys., **98** 145, 1935.

Fock, V.A., *Hydrogen atoms and non-Euclidian geometry*, Kgl. Norske Vidensskab Forh, **31** 138, 1958.

Fonseca, A.C. and Pena, M.T., *Rotational-invariant Sturmian-Faddeev ansatz for the solution of hydrogen molecular ion (H2+): a general approach to molecular three- body problems*, Phys. Rev. A: Gen. Phys., **38** (10) 4967-84, 1988.

Fonseca, A.C., *Four-body equations in momentum space*, Lect. Notes Phys., **273** (Models Methods Few-Body Phys.), p 161-200, 1987.

Fripiat, J.G., Delhalle, J. and Defranceschi, M., *A momentum space approach to improve ab initio Hartree-Fock results based on the LCAO-GTF approximation*, NATO ASI Ser, Ser C, **271** (Numer. Determ. Electron. Struct. At., Diat. Polyat. Mol.) 263-8, 1989.

Gadre, S.R. and Bendale, R.D., *Maximization of atomic information-entropy sum in configuration and momentum spaces*, Int. J. Quantum Chem., **28** (2) 311-14, 1985.

Gadre, S.R. and Chakravorty, S., *The self-interaction correction to the local spin density model: effect on atomic momentum space properties*, Chem. Phys. Lett., **120** (1) 101-5, 1985.

Gallaher, D.F. and Wilets, L., *Coupled-state calculations of proton-hydrogen scattering in the Sturmian representation*, Phys. Rev., **169** (1) 139-49, 1968.

Gaunt, J.A., Phil. Trans. Roy. Soc. Lond., **228** 151, 1929; Proc. Roy. Soc., **A222** 513, 1929.

Gazeau, J.P. and Maquet, A., *A new approach to the two-particle Schrödinger bound state problem*, J. Chem. Phys., **73** (10) 5147-54, 1980.

Gazeau, J.P. and Maquet, A., *Bound states in a Yukawa potential: a Sturmian group theoretical approach*, Phys. Rev. A, **20** (3) 727-39, 1979.

Geller, M., *Two-center Coulomb integrals*, J. Chem. Phys., **41** 4006, 1964.

Gerry, C.C., *Inner-shell bound-bound transitions from variationally scaled Sturmian functions*, Phys. Rev. A: Gen. Phys., **38** (7) 3764-5, 1988.

Ghosh, S.K., *Quantum chemistry in phase space: some current trends*, Proc. Indian Acad. Sci., Chem. Sci., **99** (1-2) 21-8, 1987.

Gloeckle W., *Few-body equations and their solutions in momentum space*, Lect. Notes Phys., **273** (Models Methods Few-Body Phys.), p 3-52, 1987.

Goscinski, O., *Preliminary Research Report No. 217*, Quantum Chemistry Group, Uppsala University, 1968.

Goscinski, O. and Lindner, P., **Natural Spin Orbitals and Generalized Overlap Amplitudes, Preliminary Research Report No. 238**, Quantum Chemistry Group, Uppsala University, 1969.

Goscinski, O., Adv. Quantum Chem., **41** 51-85, 2003.

Gradshteyn, I.S. and Ryshik, I.M., **Tables of Integrals, Series and Products**, Academic Press, New York, 1965.

Grant, I.P., in **Relativistic Effects in Atoms and Molecules**, Wilson, S., Ed., Plenum Press, 1988.

Grant, I.P., in **Atomic, Molecular and Optical Physics Handbook**, Drake, G.W.F. Ed., Chapt 22, p 287, AIP Press, Woodbury New York, 1996.

Greene, C.H., Phys. Rev., **A23** 661, 1981.

Gross, E.K.U. and Dreizler, R.M. (eds.), *Density Functional Theory*, Plenum Press, New York, 1995.

Gruzdev, P.F., Soloveva, G.S. and Sherstyuk, A.I., *Calculation of neon and argon steady-state polarizabilities by the method of Hartree-Fock SCF Sturmian expansion*, Opt. Spektrosk., **63** (6) 1394-7, 1987.

Haftel, M.I. and Mandelzweig, V.B., *A fast convergent hyperspherical expansion for the helium ground state*, Phys. Lett., **A120** 232, 1987.

Han, C.S., *Electron-atom scattering in an intense radiation field*, Phys. Rev. A: At., Mol., Opt. Phys., **51** (6) 4818-23, 1995.

Hansen, T.B., *The many-center one-electron problem in momentum space*, Thesis, Chemical Institute, University of Copenhagen, 1998.

Harris, F.E. and Michels, H.H., Adv. Chem. Phys. **13**, 205, 1967.

Hartt, K. and Yidana, P.V.A., *Analytic Sturmian functions and convergence of separable expansions*, Phys. Rev. C: Nucl. Phys., **36** (2) 475-84, 1987.

Heddle, D. P., Kwon, Y.R. and Tabakin, F., *Coulomb plus strong interaction bound states-momentum space numerical solutions*, Comput. Phys. Commun., **38** (1) 71-82, 1985.

Heller, E.J., *Wigner phase space method: Analysis for semiclassical applications*, J. Chem. Phys., **65** 1289-1298, 1976.

Henderson, G. A., *Variational theorems for the single-particle probability density and density matrix in momentum space*, Phys. Rev. A, **23** (1) 19-20, 1981.

Henriksen, N.E., Billing, G.D. and Hansen, F.Y., *Phase-space representation of quantum mechanics: Dynamics of the Morse oscillator*, Chem. Phys. Lett., **148** 397-403.

Herrick, D.R., *Variable dimensionality in the group-theoretic prediction of configuration mixings for doubly-excited helium*, J. Math. Phys., **16** 1046, 1975.

Herrick, D.R., *New symmetry properties of atoms and molecules*, Adv. Chem. Phys., **52** 1, 1983.

Herschbach, D. R., *Dimensional interpolation for two-electron atoms*, J. Chem. Phys., **84** 838, 1986.

Herschbach, D. R., Avery, J. and Goscinski, O., Eds., **Dimensional Scaling in Chemical Physics**, Kluwer, Dordrecht, 1993.

Hietschold, M., Wonn, H. and Renz, G., *Hartree-Fock-Slater exchange for anisotropic occupation in momentum space*, Czech J. Phys., **B35** (2) 168-75, 1985.

Hillery, M., O'Connell, R.F., Scully, M.O., and Wigner, E.P., *Distribution functions in physics: Fundementals*, Physics Reports, **106** 121-167, 1984.

Holoeien, E. and Midtdal, J., *Variational nonrelativistic calculations for the (2pnp)1,3Pe states of two-electron atomic systems*, J. Phys. B, **4** (10) 1243-9, 1971.

Holz, J., *Self-energy of electrons in a Coulomb field: momentum-space method*, Z. Phys. D: At., Mol. Clusters, **4** (3) 211-25, 1987.

Horacek, J. and Zejda, L., *Sturmian functions for nonlocal interactions*, Czech J. Phys., **43** (12) 1191-201, 1993.

Hughs, J.W.B., Proc. Phys. Soc., **91** 810, 1967.

Hua, L.K., **Harmonic Analysis of Functions of Several Complex Variables in the Classical Domains**, American Mathematical Society, Providence, R.I., 1963.

Ihm, J., Zunger, A. and Cohen, M. L., *Momentum-space formalism for the total energy of solids*, J. Phys. C, **12** (21) 4409-22, 1979.

Ishikawa, Y., Rodriguez, W., and Alexander, S.A., *Solution of the integral Dirac equation in momentum space*, Int. J. Quantum Chem., Quantum Chem. Symp., **21** 417-23, 1987.

Ishikawa, Y., Rodriguez, W., Torres, S. and Alexander S.A., *Solving the Dirac equation in momentum space: a numerical study of hydrogen diatomic monopositive ion*, Chem. Phys. Lett., **143** (3) 289-92, 1988.

Jain, A. and Winter, T.G., *Electron transfer, target excitation, and ionization in H+ + Na(3s) and H+ + Na(3p) collisions in the coupled-Sturmian-pseudostate approach*, Phys. Rev. A: At., Mol., Opt. Phys., **51** (4) 2963-73, 1995.

Jain, B.L., *A numerical study on the choice of basis sets used for translating ETOs in multi-center LCAO calculations*, ETO Multicent Mol Integr, Proc Int Conf, 1st, Reidel, Dordrecht, Neth, p 129-33, 81, Ed. Weatherford, Charles A. ; Jones, Herbert W., 1982.

Jolicard, G., and Billing, G.D., *Energy dependence study of vibrational inelastic collisions using the wave operator theory and an analysis of quantum flows in momentum space*, Chem. Phys., **149** (3) 261-73, 1991.

Judd, B.R., **Angular Momentum Theory for Diatomic Molecules**, Academic Press, New York, 1975.

Kaijser, P. and Lindner, P., *Momentum distribution of diatomic molecules*, Philos. Mag., **31** (4) 871-82, 1975.

Kaijser, P. and Sabin, J.R., *A comparison between the LCAO-X.alpha. and Hartree-Fock wave functions for momentum space properties of ammonia*, J. Chem. Phys., **74** (1) 559-63, 1981.

Karule, E. and Pratt, R.H., *Transformed Coulomb Green function Sturmian expansion*, J. Phys. B: At., Mol. Opt. Phys., **24** (7) 1585-91, 1991.

Katyurin, S.V. and Glinkin, O.G., *Variation-iteration method for one-dimensional two-electron systems*, Int. J. Quantum Chem., **43** (2) 251-8, 1992.

Kellman, M.E. and Herrick, D.R., *Ro-vibrational collective interpretation of supermultiplet classifications of intrashell levels of two-electron atoms*, Phys. Rev. A, **22** 1536, 1980.

Kil'dyushov, M.S., Sov. J .Nucl. Phys., **15** 113, 1972.

Kil'dyushov, M.S., and Kuznetsov, G.I., Sov. J. Nucl. Phys., **17** 1330, 1973.

King, H.F., Stanton, R.E., Kim, H., Wyatt, R.E., and Parr, R.G., J. Chem. Phys., **47** 1936, 1967.

Klar, H., J. Phys. B, **7** L436, 1974.

Klar, H. and Klar, M., *An accurate treatment of two-elecctron systems*, J. Phys. B, **13** 1057, 1980.

Klar, H., *Exact atomic wave functions - a generalized power-series expansion using hyperspherical coordinates*, J. Phys. A, **18** 1561, 1985.

Klarsfeld, S. and Maquet, A., *Analytic continuation of sturmian expansions for two-photon ionization*, Phys. Lett. A, **73A** (2) 100-2, 1979.

Klarsfeld, S. and Maquet, A., *Pade-Sturmian approach to multiphoton ionization in hydrogenlike atoms*, Phys. Lett. A, **78A** (1) 40-2, 1980.

Klepikov, N.P., Sov. J. Nucl. Phys. **19** 462, 1974.

Knirk, D.L., *Approach to the description of atoms using hyperspherical coordinates*, J. Chem. Phys., **60** 1, 1974.

Koga, T. and Murai, Takeshi, *Energy-density relations in momentum space. III. Variational aspect*, Theor. Chim. Acta, **65** (4) 311-16, 1984.

Koga, T., *Direct solution of the $H(1s) - H^+$ long-range interactionproblem in momentum space*, J. Chem. Phys., **82** 2022, 1985.

Koga, T. and Matsumoto, S., *An exact solution of the interaction problem between two ground-state hydrogen atoms*, J. Chem. Phys., **82** 5127, 1985.

Koga, T. and Kawaai, R., *One-electron diatomics in momentum space. II. Second and third iterated LCAO solutions* J. Chem. Phys., **84** (10) 5651-4, 1986.

Koga, T. and Matsuhashi, T., *One-electron diatomics in momentum space. III. Nonvariational method for single-center expansion*, J. Chem. Phys., **87** (3) 1677-80, 1987.

Koga, T. and Matsuhashi, T., *Sum rules for nuclear attraction integrals over hydrogenic orbitals*, J. Chem. Phys., **87** (8) 4696-9, 1987.

Koga, T. and Matsuhashi, T., *One-electron diatomics in momentum space. V. Nonvariational LCAO approaaach*, J. Chem. Phys., **89** 983, 1988.

Koga, T., Yamamoto, Y. and Matsuhashi, T., *One-electron diatomics in momentum space. IV. Floating single-center expansion*, J. Chem. Phys., **88** (10) 6675-6, 1988.

Koga, T. and Ougihara, T., *One-electron diatomics in momentum space. VI. Nonvariational approach to excited states*, J. Chem. Phys., **91** (2) 1092-5, 1989.

Koga, T., Horiguchi, T. and Ishikawa, Y., *One-electron diatomics in momentum space. VII. Nonvariational approach to ground and excited states of heteronuclear systems*, J. Chem. Phys., **95** (2) 1086-9, 1991.

Kolos, W. and Wolniewicz, L., J. Chem. Phys., **41** 3663, 1964.

Kolos, W. and Wolniewicz, L., J. Chem. Phys., **49** 404, 1968.

Kramer, P. J. and Chen, J.C.Y., *Faddeev equations for atomic problems. IV. Convergence of the separable-expansion method for low-energy positron-hydrogen problems*, Phys. Rev. A, **(3)3** (2) 568-73, 1971.

Krause, J.L. and Berry, R.S., *Electron correlation in alkaline earth atoms*, Phys. Rev. A, **31** (5) 3502-4, 1985.

Kristoffel, N., *Statistics with arbitrary maximal allowed number of particles in the cell of the momentum space (methodical note)*, Eesti. Tead. Akad. Toim., Fuus., Mat., **41** (3) 207-10, 1992.

Kupperman, A. and Hypes, P.G., *3-dimensional quantum mechanical reactive scattering using symmetrized hyperspherical coordinates*, J. Chem. Phys., **84** 5962, 1986.

Kuznetsov, G.I. and Smorodinskii, Ya., Sov. J. Nucl. Phys., **25** 447, 1976.

Lakshmanan, M. and Hasegawa, H., *On the canonical equivalence of the Kepler problem in coordinate and momentum spaces*, J. Phys. A: Math. Gen. **17** (16), 1984.

Lanczos, C., *An iteration method for the solution of the eigenvalue problem of linear differential and integral operators*, J. Res. Natl. Bur. Stand., **45** 255-82, 1950.

Landau, L.D., and Lifshitz, E.M., **Quantum Mechanics; Non-Relativistic Theory**, Pergamon Press, London, 1959.
Lassettre, E.N., *Momentum eigenfunctions in the complex momentum plane. V. Analytic behavior of the Schrödinger equation in the complex momentum plane. The Yukawa potential*, J. Chem. Phys., **82** (2) 827-40, 1985.
Lassettre, E.N., *Momentum eigenfunctions in the complex momentum plane. VI. A local potential function*, J. Chem. Phys., **83** (4) 1709-21, 1985.
Lin, C.D., Phys. Rev. Lett. **35** 1150, 1975.
Lin, C.D., *Analytical channel functions for 2-electron atoms in hyperspherical coordinates*, Phys. Rev. A, **23** 1585, 1981.
Linderberg, J. and Öhrn, Y., *Kinetic energy functional in hyperspherical coordinates*, Int. J. Quantum Chem., **27** 273, 1985.
Liu, F.Q., Hou, X.J. and Lim, T.K., *Faddeev-Yakubovsky theory for four-body systems with three-body forces and its one-dimensional integral equations from the hyperspherical-harmonics expansion in momentum space*, Few-Body Syst., **4** (2) 89-101, 1988.
Liu, F.Q. and Lim, T.K., *The hyperspherical-harmonics expansion method and the integral-equation approach to solving the few-body problem in momentum space*, Few-Body Syst., **5** (1) 31-43, 1988.
Lizengevich, A.I., *Momentum correlations in a system of interacting particles*, Ukr Fiz Zh (Russ Ed), **33** (10) 1588-91, 1988.
López, R., Ramírez, G., García de la Vega, J.M., and Fernández Rico, J., J. Chim. Phys., **84** 695, 1987.
Louck, J.D., *Generalized orbital angular momentum and the n-fold degenerate quantum mechanical oscillator*, J. Mol. Spectr., **4** 298, 1960.
Louck, J.D. and Galbraith, H.W., Rev. Mod. Phys., **44** 540, 1972.
Löwdin, P.O., Phys. Rev. **97** 1474, 1955.
Löwdin, P.O., Appl. Phys. Suppl., **33** 251, 1962.
Macek, J., J. Phys. **B1** 831, 1968.
McWeeny, R., and Coulson, C.A., *The computation of wave functions in momentum space. I. The helium atom*, Proc. Phys. Soc. (London) A, **62** 509, 1949.
McWeeny, R., *The computation of wave functions in momentum space. II. The hydrogen molecule ion*, Proc. Phys. Soc. (London) A, **62** 509, 1949.
McWeeny, R., **Methods of Molecular Quantum Mechanics**, (second edition), 1989.
Manakov, N.L., Rapoport, L.P. and Zapryagaev, S.A., *Sturmian expansions of the relativistic Coulomb Green function*, Phys. Lett. A, **43** (2) 139-40, 1973.
Maquet, Alfred, M., Philippe and Veniard, Valerie, *On the Coulomb Sturmian basis*, NATO ASI Ser, Ser C, **271** (Numer. Determ. Electron. Struct. At., Diat. Polyat. Mol.), 295-9, 1989.
Maruani, Jean, editor, **Molecules in Physics, Chemistry and Biology**, 3, Kluwer Academic Publishers, Dordrecht, 1989.
McCarthy, I.E. and Rossi, A.M., *Momentum-space calculation of electron-molecule scattering*, Phys. Rev. A: At., Mol., Opt. Phys., **49** (6) 4645-52, 1994.

McCarthy, I.E. and Stelbovics, A.T., *The momentum-space coupled-channels-optical method for electron-atom scattering*, Flinders Univ. South Aust., Inst. At. Stud., (Tech Rep) FIAS-R, (FIAS- R-111,) 51 pp., 1983.

Michels, M.A.J., Int. J. Quantum Chem., **20** 951, 1981.

Mizuno, J., *Use of the Sturmian function for the calculation of the third harmonic generation coefficient of the hydrogen atom*, J. Phys. B, **5** (6) 1149-54, 1972.

Monkhorst, H.J. and Harris, F.E., *Accurate calculation of Fourier transform of two-center Slater orbital products*, Int. J. Quantum Chem., **6** 601, 1972.

Monkhorst, H.J. and Jeziorski, B., *No linear dependence or many-center integral problems in momentum space quantum chemistry*, J. Chem. Phys., **71** (12) 5268-9, 1979.

Moore, C.E., **Atomic Energy Levels; Circular of the National Bureau of Standards 467**, Superintendent of Documents, U.S. Government Printing Office, Washington 25 D.C., 1949.

National Institute for Fusion Science (NIFS), *Atomic and Molecular Databases: Data for Autoionizing States*, http://dprose.nifs.ac.jp/DB/Auto/

National Institute of Standards and Technology (NIST), *NIST Atomic Spectra Database*, http://physics.nist.gov/asd

National Institute of Standards and Technology (NIST), *CODATA Recommended Values of the Fundamental Physical Constants: 2002*, http://physics.nist.gov/cuu/Constants/

Navasa, J. and Tsoucaris, G., *Molecular wave functions in momentum space*, Phys. Rev. A, **24** 683, 1981.

Nikiforov, A.F., Suslov, S.K. and Uvarov, V.B., **Classical Orthogonal Polynomials of a Discrete Variable**, Springer-Verlag, Berlin, 1991.

Norbury, J.W., Maung, K.M. and Kahana, D.E., *Exact numerical solution of the spinless Salpeter equation for the Coulomb potential in momentum space*, Phys. Rev. A: At., Mol., Opt. Phys., **50** (5) 3609-13, 1994.

Novosadov, B.K., Opt. Spectrosc., **41** 490, 1976.

Novosadov, B.K., Int. J. Quantum Chem., **24** 1, 1983.

Ojha, P.C., *The Jacobi-matrix method in parabolic coordinates: expansion of Coulomb functions in parabolic Sturmians*, J. Math. Phys. (N Y), **28** (2) 392-6, 1987.

Park, I.H., Kim, H.J. and Kang, J.S., *Computer simulation of quantum mechanical scattering in coordinate and momentum space* Sae. Mulli., **26** (4) 155-67, 1986.

Pathak, R.K., Kulkarni, S.A. and Gadre, S.R., *Momentum space atomic first-order density matrices and "exchange-only" correlation factors*, Phys. Rev. A, **42** (5) 2622-6, 1990.

Pathak, Rajeev K., Panat, Padmakar V. and Gadre, S.R., *Local-density-functional model for atoms in momentum space*, Phys. Rev. A, **26** (6) 3073-7, 1982.

Pauling, L., and Wilson, E.B., **Introduction to Quantum Mechanics**, McGraw-Hill, 1935.

Pisani, L. and Clementi, E., in **Methods and Techniques in Computational Chemistry**, Clementi, E., and Corongiu, G., Eds., STEF, Cagliari, 1995.

Plante, D.R., Johnson, W.R., and Sapirstein, J. Phys. Rev. **A49** 3519, 1994.

Podolski, B., Proc. Natl. Acad. Sci. (USA), **14** 253, 1928.

Podolski, B. and Pauling, L., Phys. Rev., **34** 109, 1929.

Potvliege, R.M. and Shakeshaft, R., *Determination of the scattering matrix by use of the Sturmian representation of the wave function: choice of basis wave number*, J. Phys. B: At., Mol. Opt. Phys., **21** (21) L645, 1988.

Potvliege, R.M. and Smith, Philip H.G., *Stabilization of excited states and harmonic generation: Recent theoretical results in the Sturmian-Floquet approach*, NATO ASI Ser, Ser B, **316** (Super-Intense Laser-Atom Physics) 173-84, 1993.

Pyykkö, P., **Relativistic Theory of Atoms and Molecules. A Bibliography, 1916-1985**, Lecture Notes in Chemistry, **41**, 1986.

Pyykkö, P., Chem. Rev., **88** 563, 1988.

Rahman, N.K., *On the Sturmian representation of the Coulomb Green's function in perturbation calculation*, J. Chem. Phys., **67** (4) 1684-5, 1977.

Rawitscher, G.H. and Delic, G., *Sturmian representation of the optical model potential due to coupling to inelastic channels*, Phys. Rev. C, **29** (4) 1153-62, 1984.

Rawitscher, G. H. and Delic, G., *Solution of the scattering T matrix equation in discrete complex momentum space*, Phys. Rev. C, **29** (3) 747-54, 1984.

Read, F.H., *A new class of atomic states: The 'Wannier-ridge' resonances*, Aus. J. Phys. **35** 475-99, 1982.

Regier, P.E., Fisher, J., Sharma, B.S. and Thakkar, A.J., *Gaussian vs. Slater representations of d orbitals: An information theoretic appraisal based on both position and momentum space properties*, Int. J. Quantum Chem., **28** (4) 429-49, 1985.

Rettrup, S., **Lecture Notes on Quantum Chemistry**, Department of Chemistry, University of Copenhagen, 2003.

Ritchie, B., *Comment on "Electron molecule scattering in momentum space"*, J. Chem. Phys., **72** (2) 1420-1, 1980.

Ritchie, B., *Electron-molecule scattering in momentum space*, J. Chem. Phys., **70** (6) 2663-9, 1979.

Rodriguez, W. and Ishikawa, Y., *Fully numerical solutions of the Hartree-Fock equation in momentum space: a numerical study of the helium atom and the hydrogen diatomic monopositive ion*, Int. J. Quantum Chem., Quantum Chem. Symp., **22** 445-56, 1988.

Rodriguez, W. and Ishikawa, Y., *Fully numerical solutions of the molecular Schrödinger equation in momentum space*, Chem. Phys. Lett., **146** (6) 515-17, 1988.

Rohwedder, B. and Englert, B.G., *Semiclassical quantization in momentum space*, Phys. Rev. A: At., Mol., Opt. Phys., **49** (4) 2340-6, 1994.

Rotenberg, M., Ann. Phys. (New York), **19** 262, 1962.

Rotenberg, M., *Theory and application of Sturmian functions*, Adv. At. Mol. Phys., **6** 233-68, 1970.

Royer, A., *Wigner function as the expectation value of a parity operator*, Phys. Rev. A, **15** 449-450, 1977.

Rudin, W., **Fourier Analysis on Groups**, Interscience, New York, 1962.

Schmider, H., Smith, V.H. Jr. and Weyrich, W., *On the inference of the one-particle density matrix from position and momentum-space form factors*, Z. Naturforsch., A: Phys. Sci., **48** (1-2) 211-20, 1993.

Schmider, H., Smith, V.H. Jr. and Weyrich, W., *Reconstruction of the one-particle density matrix from expectation values in position and momentum space*, J. Chem. Phys., **96** (12) 8986-94, 1992.

Schuch, D., *On a form of nonlinear dissipative wave mechanics valid in position- and momentum-space*, Int. J. Quantum Chem., Quantum Chem. Symp., **28** (Proceedings of the International Symposium on Atomic, Molecular, and Condensed Matter Theory and Computational Methods, 1994) 251-9, 1994.

Shabaev, V.M., *Relativistic Coulomb Green function with regard to finite size of the nucleus*, Vestn. Leningr. Univ., Fiz., Khim. (2), p 92-6, 1984.

Shakeshaft, R. and Tang, X., *Determination of the scattering matrix by use of the Sturmian representation of the wave function*, Phys. Rev. A: Gen. Phys., **35** (9) 3945-8, 1987.

Shakeshaft, R., *A note on the Sturmian expansion of the Coulomb Green's function*. J. Phys. B: At. Mol. Phys., **18** (17) L611-L615, 1985.

Shakeshaft, R., *Application of the Sturmian expansion to multiphoton absorption: hydrogen above the ionization threshold*, Phys. Rev. A: Gen. Phys., **34** (6) 5119-22, 1986.

Shakeshaft, R., *Coupled-state calculations of proton-hydrogen-atom scattering with a Sturmian expansion*, Phys. Rev. A, **14** (5) 1626-33, 1976.

Shakeshaft, R., *Sturmian expansion of Green's function and its application to multiphoton ionization of hydrogen*, Phys. Rev. A: Gen. Phys., **34** (1) 244-52, 1986.

Shakeshaft, R., *Sturmian basis functions in the coupled state impact parameter method for hydrogen(+) + atomic hydrogen scattering*, J. Phys. B, **8** (7) 1114-28, 1975.

Shelton, D.P., *Hyperpolarizability of the hydrogen atom*, Phys. Rev. A: Gen. Phys., **36** (7) 3032-41, 1987.

Sherstyuk, A.I., *Sturmian expansions in the many-fermion problem*, Teor. Mat. Fiz., **56** (2) 272-87, 1983.

Shibuya, T. and Wulfman, C.E., *Molecular orbitals in momentum space*, Proc. Roy. Soc. A, **286** 376, 1965.

Shull, H. and Löwdin, P.-O., *Superposition of configurations and natural spin-orbitals. Applications to the He problem*, J. Chem. Phys., **30** 617, 1959

Simas, A.M., Thakkar, A.J. and Smith, V.H. Jr., *Momentum space properties of various orbital basis sets used in quantum chemical calculations*, Int. J. Quantum Chem., **21** (2) 419-29, 1982.

Sloan, I.H. and Gray, J.D., *Separable expansions of the t-matrix*, Phys. Lett. B, **44** (4) 354-6, 1973.

Sloan, Ian H., *Sturmian expansion of the Coulomb t matrix*, Phys. Rev. A, **7** (3) 1016-23, 1973.

Smirnov, Yu. F. and Shitikova, K.V., Sov. J. Part. Nucl., **8** 344, 1976.

Smith, F.T., *Generalized angular momentum in many-body collisions*, Phys. Rev., **120** 1058, 1960.

Smith, F.T., *A symmetric representation for three-body problems. I. Motion in a plane*, J. Math. Phys., **3** 735, 1962.
Smith, F.T., *Participation of vibration in exchange reactions*, J. Chem. Phys., **31** 1352-9, 1959.
Smith, V.H. Jr., *Density functional theory and local potential approximations from momentum space considerations*, Local Density Approximations Quantum Chem. Solid State Phys., (Proc. Symp.), Plenum, New York, N. Y, p 1-19, 82, Eds. Dahl, J.P.; Avery, J., 1984.
Smorodinskii, Ya., and Efros, V.D., Sov. J. Nucl. Phys. **17** 210, 1973.
Springborg, M. and Dahl, J.P., *Wigner's phase-space function and atomic structure*, Phys. Rev. A, **36** 1050-62, 1987.
Szmytkowski, R., *The Dirac-Coulomb Sturmians and the series expansion of the Dirac-Coulomb Green function; Application to the relativistic polarizability of the hydrogenlike atom*, J. Phys. A, **31** 4963, 1998.
Szmytkowski, R., *The continuum Schrödinger-Coulomb and Dirac-Coulomb Sturmian functions*, J. Phys. A, **31** 4963, 1998.
Szmytkowski, R., *The continuum Schrödinger-Coulomb and Dirac-Coulomb Sturmian functions*, J. Phys. A, **31** 4963, 1998.
Taieb, R.; Veniard, V.; Maquet, A.; Vucic S. and Potvliege R.M., *Light polarization effects in laser-assisted electron-impact-ionization ((e,2e)) collisions: a Sturmian approach*, J. Phys. B: At., Mol. Opt. Phys., **24** (14) 3229-40, 1991.
Tang, X. and Shakeshaft, R., *A note on the solution of the Schrödinger equation in momentum space*, Z. Phys. D: At., Mol. Clusters, **6** (2) 113-17, 1987.
Tarter, C.B., J. Math. Phys., **11** 3192, 1970.
Tel-nov, D.A., *The d.c. Stark effect in a hydrogen atom via Sturmian expansions*, J. Phys. B: At., Mol. Opt. Phys., **22** (14) L399-L404, 1989.
Thakkar, A.J. and Koga, T., *Analytic approximations to the momentum moments of neutral atoms*, Int. J. Quantum Chem., Quantum Chem. Symp., **26** (Proc. Int. Symp. At., Mol., Condens. Matter Theory Comput. Methods, 1992) 291-8, 1992.
Thakkar, A.J. and Tatewaki, H., *Momentum-space properties of nitrogen: improved configuration-interaction calculations*, Phys. Rev. A, **42** (3) 1336-45, 1990.
Tzara, C., *A study of the relativistic Coulomb problem in momentum space*, Phys. Lett. A, **111A** (7) 343-8, 1985.
Ugalde, J.M., *Exchange-correlation effects in momentum space for atoms: an analysis of the isoelectronic series of lithium 2S and beryllium 1S*, J. Phys. B: At. Mol. Phys., **20** (10) 2153-63, 1987.
Vainshtein, L.A. and Safronova, V.I., *Wavelengths and transition probabilities of satellites to resonance lines of H- and He-like ions*, Atomic Data and Nucl.Data Tables, *21*, 49, 1978.
Vainshtein, L.A. and Safronova, V.I., *Dielectronic satellite spectra for highy charged H-like ions and He-like ions with Z=6-33*, Atomic Data and Nucl. Data Tables, **25** 311, 1980.
Van Haeringen, H. and Kok, L.P., *Inequalities for and zeros of the Coulomb*

T matrix in momentum space, Few Body Probl Phys, Proc Int IUPAP Conf, 10th, North-Holland, Amsterdam, Neth, p 361-2, 83, Ed. Zeitnitz, Bernhard, 1984.

Vilenkin, N.K., *Special Functions and the Theory of Group Representations*, American Mathematical Society, Providence, R.I., 1968.

Vilenkin, N. Ya., Kuznetsov, G.I., and Smorodinskii, Ya.A., Sov. J. Nucl. Phys., **2** 645, 1966.

Vladimirov, Yu.S. and Kislov, V.V., *Charge of the nucleus of a hydrogen-like atom as an eigenvalue of a 6-dimensional wave equation in momentum space*, Izv. Vyssh. Uchebn. Zaved., Fiz., **28** (4) 66-9, 1985.

Weatherford, C.A., *Scaled hydrogenic Sturmians as ETOs*, ETO Multicent. Mol. Integr., Proc. Int. Conf., 1st, Reidel, Dordrecht, Neth., p 29-34, 81, Ed. Weatherford, Charles A. ; Jones, Herbert W., 1982.

Wen, Z.-Y. and Avery, J., *Some properties of hyperspherical harmonics*, J. Math. Phys., **26** 396, 1985.

Weniger, E.J., *Weakly convergent expansions of a plane wave and their use in Fourier integrals*, J. Math. Phys., **26** 276, 1985.

Weniger, E.J. and Steinborn, E.O., *The Fourier transforms of some exponential-type basis functions and their relevance for multicenter problems*, J. Chem. Phys., **78** 6121, 1983.

Weniger, E.J., Grotendorst, J., and Steinborn, E.O., *Unified analytical treatment of overlap, two-center nuclear attraction, and Coulomb integrals of B functions via the Fourier transform method*, Phys. Rev. A, **33** 3688, 1986.

Whitten, R.C. and Sims, J.S., Phys. Rev. A, **9** 1586, 1974.

Wigner, E., Phys. Rev., **40** 749, 1932.

Windt, L. de, Fischer, P., Defranceschi, M., Delhalle, J. and Fripiat, J. G., *A combined analytical and numerical strategy to solve the atomic Hartree-Fock equations in momentum space*, J. Comput. Phys., **111** (2) 266-74, 1994.

Winter, T. G. and Alston, S.G., *Coupled-Sturmian and perturbative treatments of electron transfer and ionization in high-energy helium p-He+ collisions*, Phys. Rev. A, **45** (3) 1562-8, 1992.

Winter, T.G., *Electron transfer and ionization in collisions between protons and the ions lithium(2+) and helium(1+) studied with the use of a Sturmian basis*, Phys. Rev. A: Gen. Phys., **33** (6) 3842-52, 1986.

Winter, T.G., *Coupled-Sturmian treatment of electron transfer and ionization in proton-neon collisions*, Phys. Rev. A, **48** (5) 3706-13, 1993.

Winter, T.G., *Electron transfer and ionization in collisions between protons and the ions helium(1+), lithium(2+), beryllium(3+), boron(4+), and carbon(5+) studied with the use of a Sturmian basis*, Phys. Rev. A: Gen. Phys., **35** (9) 3799-809, 1987.

Winter, T.G., *Electron transfer in p-helium(1+) ion and helium(2+) ion-atomic helium collisions using a Sturmian basis*, Phys. Rev. A, **25** (2) 697-712, 1982.

Winter, T.G., *Sturmian treatment of excitation and ionization in high-energy proton-helium collisions*, Phys. Rev. A, **43** (9) 4727-35, 1991.

Winter, T.G., *Coupled-Sturmian treatment of electron transfer and ionization in proton-carbon collisions*, Phys. Rev. A, **47** (1) 264-72, 1993.

Winter, T.G., *Electron transfer and ionization in proton-helium collisions studied using a Sturmian basis*, Phys. Rev. A, **44** (7) 4353-67, 1991.

Wulfman, C.E., *Semiquantitative united-atom treatment and the shape of triatomic molecules*, J. Chem. Phys. **31** 381, 1959.

Wulfman, C.E., *Dynamical groups in atomic and molecular physics*, in **Group Theory and its Applications**, Loebel, E.M. Ed., Academic Press, 1971.

Wulfman, C.E., *Approximate dynamical symmetry of two-electron atoms*, Chem. Phys. Lett.ers **23** (3), 1973.

Wulfman, C.E., *On the space of eigenvectors in molecular quantum mechanics*, Int. J. Quantum Chem. **49** 185, 1994.

Yurtsever, E.; Yilmaz, O., and Shillady, D.D., *Sturmian basis matrix solution of vibrational potentials*. Chem. Phys. Lett., **85** (1) 111-16, 1982.

Yurtsever, E., *Franck-Condon integrals over a sturmian basis. An application to photoelectron spectra of molecular hydrogen and molecular nitrogen*, Chem. Phys. Lett., **91** (1) 21-6, 1982.

Index

Absolute velocity, 79
Additional indices, 130
Alternative 4-dimensional
 hyperspherical harmonics, 115, 139
Alternative chains of subgroups, 131
Alternative hyperspherical harmonics,
 139, 141
Alternative representations, 196
Amos, A.T., 175
Angular correlation, 47
Angular integrals, 168, 170
Angular integration formula, 126
Angular momentum, 48
Angular momentum quantum
 number, 170
Anomalous states, 70
Anticommutation, 82
Antisymmetric functions, 13, 203
Antisymmetrizer, 176, 203
Appropriate domains for iteration, 45
Approximate N-electron Dirac
 equation, 91
Approximate N-electron Schrödinger
 equation, 20, 22, 25
Aquilanti, V., xiv, 131, 139, 156
Atomic calculations, 20
Atomic spectra, 19
Atomic units, 5, 6, 39, 69, 76, 81, 82,
 187, 193
Autoionizing states, 30, 32, 34, 48
Automatic selection of basis sets, 16,
 44

Avery, J.S., 136, 182, 184
Azimuthal quantum numbers, 44

Basis functions, 19, 25
Basis-set-free iteration, 17
Berry, S., 47
Berylliumlike isoelectronic series, 35
Bessel functions, 108
Bessel functions, modified, 198
Bessel functions, spherical, 198
Binding energy, 32, 91
Binomial coefficients, 34, 48
Boronlike isoelectronic series, 35
Boundary conditions, 1, 203

Calculated energies, 51
Caligiana, A., xiv, 131, 156
Canonical decomposition of
 homogeneous polynomials, 119,
 121, 124, 128, 140
Carbonlike isoelectronic series, 34, 35
Cartesian coordinates, 19, 125
Chains of subgroups, 111, 130, 131,
 141
Charge conservation, 81
Charge density, 80
Choice of domains, 45
Clebsch-Gordan coefficients, 26, 32,
 84, 171
Clementi, E., 58
Coletti, C., xiv, 131, 139
Commutation, 195

Complete canonical decomposition, 122
Completeness, xi, 13, 15, 16, 44, 112, 202–204
Confluent hypergeometric function, 3, 21, 85
Conjugate eigenvalue problem, 1, 3
Conjugate function, 25
Conservation of symmetry, 194, 195
Construction of hyperspherical harmonics, 131
Continuity, 1
Continuum, 2
Contour integration, 185, 198
Convergence, 17
Convergence factor, 188
Convergent series, 125, 126
Core ionization, 57, 58
Core ionization energy, 60
Core-excited states, 58
Core-ionized states, 61
Correlation, angular, 47
Coulomb Sturmian basis sets, xi
Coulomb Sturmian configurations, 201, 206
Coulomb Sturmian spin-orbitals, 201
Coulomb Sturmians, 3, 6, 107, 108, 110, 153, 155, 164
Current density 3-vector, 80
Current density 4-vector, 80, 81

D'Alembertian operator, 80
Davidson, E., 17
Decomposition of monomials, 127
Degeneracy, 31, 32, 43, 48, 111
Degenerate configurations, 31
Degree of a harmonic polynomial, 119
Degree of a hyperspherical harmonic, 129
Degree of a monomial, 117
Degree of a polynomial, 118, 171
Density distribution, 60
Derivation of the secular equations, 24
Diatomic molecules, 159
Differentiability, 1

Dipole moment, 75
Dirac alpha matrices, 82
Dirac configurations, 91–93
Dirac delta function, 10, 194
Dirac equation, 76, 81, 82, 172
Dirac equation for a hydrogenlike atom, 83
Dirac gamma matrices, 82
Dirac Hamiltonian, 83
Dirac spinors, 82
Dirac, P.A.M., 81
Direct-space solutions, 141
Displaced coordinate system, 181
Displaced function, 182
Distributions, 16
Domains, 45
Dominant nuclear attraction, 31
Double Fourier transformation, 195
Double minors, 93, 178
Doubly-excited states, 27, 30, 32, 34, 47
Dynamical symmetry, 111

Effective charge, 22, 94
Effective nuclear charge, 175
Eigenfunctions, 1
Eigenfunctions of generalized angular momentum, 129
Eigenvalues, 1
Eigenvalues of generalized angular momentum, 129
Eigenvalues of the interelectron repulsion matrix, 36, 103
Eigenvectors of the interelectron repulsion matrix, 43
Einstein, A., 79
Electric field strength, 69
Electric field vector, 80
Electric fields, 69
Electromagnetic potential 4-vector, 80, 81
Electromagnetic theory, 79
Electron exchange, 13
Electron rest mass, 82
Electrostatic potential, 79
Energy, 191

INDEX

Energy-independent matrix, 25
Equivalence of inertial frames, 79
Excited states, 26
Expansion about another center, 181
Expansion in double minors, 178
Expansion in terms of minors, 177
Expansion of a plane wave, 155
Expansion of the kernel, 190
Experimental energies, 27, 51
Experimental values, 27, 34, 38
Exponential factor, 154, 201
Exponential-type orbitals, 163
External electric fields, 69
External electromagnetic potential, 81–83
External fields, 44
External magnetic fields, 76
External potential, 32

Few-electron atoms, 26, 44
Field-point, 81
Filling of shells, 60
Fine structure constant, 82
First-iterated solution, 15, 44
Fluorinelike isoelectronic series, 38
Fock projection, 155
Fock transformation, 107, 117
Fock's projection, 110, 140, 187, 188
Fock's treatment of hydrogenlike atoms, 188
Fock, V., 110, 154, 187, 191
Four-vectors, 79
Fourier convolution theorem, 12, 14
Fourier transforms, 9, 11, 15, 108, 110, 117, 141, 154–156, 181, 187, 194, 195
Fourier-transformed hydrogenlike orbitals, 114, 187

Gamma functions, 125, 169
Gaussian basis functions, 163
Gegenbauer polynomials, 111, 134–138, 151, 181, 184, 190, 197
General fork, 142
Generalized angular momentum, 122, 130, 131, 138

Generalized Helmholtz operator, 193
Generalized Laplace equation, 119
Generalized Laplacian operator, 118, 119, 134, 140, 193
Generalized Shibuya-Wulfman integrals, 206
Generalized Slater-Condon rules, 93, 175
Generalized solid angle, 124, 125, 137, 183, 189, 196
Generalized Sturmian configurations, 76
Generalized Sturmian method, 19, 26, 70, 90
Generalized Sturmian secular equations, 24, 25
Generalized Sturmians, 5, 6, 11, 12, 16, 19, 161
Generating function, 134
Goscinski, O., xiv, 5, 20
Goscinskian configurations, 6, 19, 22–24, 26, 44, 48, 61, 76, 91, 175, 201, 204
Graphical representations, 141
Green's function, 15, 134, 193, 194, 196, 199
Green's function iteration, 16, 195
Green's function, alternative form, 15
Greene, C.H., 47
Ground states, 35, 38
Ground-state energies, 57
Group theory, 130, 194

Høloien, E., 2
Hall, G.G., 175
Hamiltonian, 25
Harmonic component, 170
Harmonic polynomial of highest degree, 120, 121, 131
Harmonic polynomials, 117–120, 122, 129, 133, 170, 191
Harmonic polynomials as eigenfunctions of generalized angular momentum, 124, 129
Harmonic projection, 131, 170, 171, 191

Hartree-Fock approximation, 58
Hartrees, 39, 91
Harvard, University, xiv
Heliumlike isoelectronic series, 27, 32, 34, 38, 51, 206
Helmholtz operator, generalized, 193
Hermiticity, 4, 23, 124, 131
Herschbach, D.R., xiv
Higher series of autoionizing states, 51
Highly-excited states, 34
Hilbert space, 2, 203
Homogeneous polynomials, 117–119, 125, 131, 139, 170
Hua, L.K., 136
Hybridization, 31
Hydrogenlike atoms, 172, 188, 191
Hydrogenlike Dirac spinors, 86
Hydrogenlike orbitals, 2–4, 107, 114, 140
Hydrogenlike Schrödinger equation, 4
Hydrogenlike spin-orbitals, 20, 24, 201
Hydrogenlike Sturmians, 117
Hydrogenlike wave equation, 22
Hyperangular integration, 124, 156
Hyperangular integration formula, 125, 126
Hypergeometric functions, 21, 85, 169
Hyperradial integral, 125
Hyperradius, 118, 124, 129, 181, 195
Hypersphere, 110, 111, 188, 189
Hyperspherical Bessel function, 136, 182
Hyperspherical expansion of a plane wave, 136
Hyperspherical harmonics, 110, 112, 117, 118, 122, 129, 131, 136, 138, 147, 149, 155, 183, 190
Hyperspherical harmonics, 4-dimensional, 134
Hyperspherical harmonics, sum rule, 137, 183

Idempotent, 195, 203
Incomplete gamma functions, 169

Induced transition dipole moment, 75
Inertial frames, 79
Infinitesimal external potential, 32
Inner product, 1, 11
Integral equation, 9, 12, 14, 187, 190, 194
Integral form of the Schrödinger equation, 194
Integral representation of the Green's function, 197
Interelectron interaction operator, 93
Interelectron repulsion, 24, 162, 178
Interelectron repulsion integrals, 163, 168
Interelectron repulsion matrix, 25, 27, 31, 33, 43, 48, 49, 57, 94
Interpolation, 94
Ionization, 47
Ions, 27
Irreducible representations, 16, 44, 130, 194
Isoelectronic series, 26, 32, 57, 61, 70
Isoenergetic basis sets, 5, 11, 16, 20
Isoenergetic solutions, xi
Isonuclear series, 60, 61
Iterated solution, 196
Iteration, 9, 14, 61, 195
Iteration using a basis, 16

Jacobi polynomials, 142

Kernel, 190
Kinetic energy operator, 19, 22
Kinetic energy term, 7
King, H.F. et al., 175
Klar, H., 47
Klar, M., 47
Koga, T., 165

Löwdin, P.O., 2, 175
Laguerre polynomials, associated, 141
Lanczos, C., 17
Laplacian operator, 4, 80, 81, 118, 134, 171
Laplacian operator, generalized, 193
Large and small components, 84, 86

Large-Z approximation, 31, 32, 34, 35, 48, 57, 204
Legendre polynomials, 111, 134, 135, 138, 168, 172, 184, 198
Lin, C.D., 47
Linearly independent harmonics, 130
Liouville, J., 1
Lithiumlike isoelectronic series, 35, 38, 45
Lorentz gauge, 81
Lorentz invariance, 79
Lorentz transformation, 79

Macek, J., 47
Magnetic field vector, 80
Magnetic fields, 76
Magnetic quantum number, 141
Major quantum numbers, 5
Many-center potentials, 158
Many-center problems, 153, 154
Many-particle Dirac equation, 92
Many-particle problems, 5
Maple, 126
Mathematica, 126, 169
Matsuhashi, T., 165
Maxwell's equations, 81
McWeeny, R., 175
Method of trees, 139, 141
Michaelson-Morley experiment, 79
Midtdal, J., 2
Minkowski space, 79, 81
Minor indices, 5, 130, 141
Minors, 177, 178
Modified Bessel function of the second kind, 197
Modified spherical Bessel differential equation, 199
Modified spherical Bessel functions of the first and second kind, 186, 199
Molecular ions, 161
Molecular problems, 206
Molecular spectra, 161, 163
Molecular wave functions, 207
Molecules, 153, 161
Momentum space, 9, 107, 110, 112, 194

Momentum space volume element, 196
Momentum-space angles, 188
Momentum-space basis sets, 108
Momentum-space hyperangles, 182
Momentum-space hyperradius, 182
Momentum-space methods, 154
Momentum-space orthonormality, 11, 12, 111
Momentum-space Schrödinger equation, 10, 14, 187, 188
Momentum-space volume element, 155, 183, 189
Monomials, 117
Motion of the nucleus, 24
Multiplicity, 45

Neonlike isoelectronic series, 38
NIST tables, 27, 38
Nitrogenlike isoelectronic series, 38
Nodes, 2
Non-degenerate roots, 43
Non-relativistic approximation, 76
Non-relativistic case, 19, 38, 93, 169, 171
Non-relativistic wave equation, 9, 24
Non-weighted orthonormality, 15
Normalization, 2, 6, 23, 25, 92, 131, 133, 161
Normalizing constant, 86, 91, 111, 135
Nuclear attraction, 24, 162, 204
Nuclear attraction integrals, 159
Nuclear attraction matrix, 25, 27
Nuclear attraction matrix diagonal and energy-independent, 25
Nuclear attraction potential, 20, 157
Nuclear charge, 32, 48, 188
Nuclear charge, weighted, 19, 21

One-electron operators, 175, 177
One-electron wave function, 187
Optimal basis set, 25
Orbital angular momentum, 123, 130
Orthogonality, 1, 124
Orthonormal basis, 2

Orthonormal sets, 111
Orthonormality, 4, 22, 92, 175
Orthonormality of hyperspherical harmonics, 131, 190
Orthonormality relation, 86, 112, 131, 136, 190
Oscillatory integral, 188
Overlap integrals, 93
Oxygenlike isoelectronic series, 38

Parabolic coordinates, 140, 149
Parity, 45, 171
Paschen-Back effect, 76
Pauli principle, 34
Pauli spin matrices, 82, 172
Permutations, 176
Perugia, University of, xiv, 139
Piecewise linearity, 60
Planck's constant, 82
Plane wave, 10, 108, 111, 188
Plane wave, d-dimensional, 182
Plane wave, hyperspherical expansion, 136, 182
Plane wave, Sturmian expansion, 13
Polarizabilities, 75
Polynomials, 126
Position vector, 79
Potential, 24
Potential at the nucleus, 61
Potential energy matrix, 16
Potential-weighted orthogonality, 5
Potential-weighted orthonormality, 4, 5, 7, 11, 22, 23, 25, 91, 93, 108, 161, 202
Primitive configurations, 27, 32, 43, 44
Principal quantum number, 24, 31, 34, 141
Probability density, 47
Projection of momentum space, 188
Projection operator, 13, 138, 191, 194, 196
Pseudo-Euclidean space-time continuum, 79
Pseudo-rotation, 79

Quantum numbers, 24, 26, 27, 110

R-block, definition, 31
R-blocks, 32, 34, 38
Radial density distribution, 60
Radial functions, 3, 107, 153
Radial functions, relativistic, 84
Radial integrals, 168, 169
Radial orthonormality, 175
Read, F.H., 47
Recursion relations, 109
Relativistic 1-electron energies, 85
Relativistic case, 171
Relativistic effects, 27, 34, 38, 79, 90
Relativistic electrodynamics, 80
Relativistic hydrogenlike orbitals, 84
Relativistic integrals, 172
Relativistic many-electron Sturmians, 90
Relativistic radial functions, 86
Relativistic secular equations, 94
Relativistic wave equation, 81
Rest energy, 82, 91
Rettrup, S., 175
Roots of the interelectron repulsion matrix, 32, 38
Rotation group, 130
Rotational modes, 47
Rotenberg, M., 2
Russell-Saunders basis sets, 26, 32, 43, 44

Scalar products between configurations, 176
Scaling parameter, 7, 24, 25, 93
Schrödinger equation, 6, 108, 193, 198
Schrödinger equation in momentum space, 187
Schrödinger equation, integral form, 14, 194
Second-iterated solution, 15, 44
Secular equations, 24, 25, 69, 94, 162, 202, 203
Separability, 20
Shavitt, I., 17
Shells, filling of, 60

INDEX

Shibuya, T., 154
Shibuya-Wulfman integrals, 154, 156, 163, 164, 203
Shifted roots, 31
Shull, H., 2
Simultaneous eigenfunctions, 32, 35, 132, 133
Simultaneous equations, 120
Singlet states, 27
Singly-excited states, 27
Slater determinant, 20, 91, 161, 175, 201
Slater exponents, 26
Slater-Condon rules, 23, 92
Slater-Condon rules, generalized, 175
Small blocks, 43
Sobolev spaces, 6, 12, 13
Solid angle, 112, 137
Solid angle, generalized, 189
Source-point, 81
Space, d-dimensional, 181
Space-component, 80
Space-time symmetry, 79–81
Special theory of relativity, 79
Spectra of few-electron atoms, 44
Spectroscopic tables, 38
Spectrum of energies, 25
Spherical Bessel differential equation, 199
Spherical Bessel functions, 108, 137, 138, 184, 188, 198
Spherical harmonics, 3, 21, 84, 108, 110, 130, 132, 135, 138, 153, 188
Spherical harmonics, sum rule, 184
Spherical polar coordinates, 133
Spherical spinor, 84
Spherical spinors, 172
Spherical symmetry, 44
Spherically averaged density, 61
Spin, 76
Spin-orbit coupling, 24
Spin-orbitals, 48, 175
Spin-spin coupling, 24
Spinor indices, 172
Spinors, 4-component, 82
Standard set, 111, 131, 141

Standard tree, 141
Stark effect, 70
Static polarizabilities, 76
Strong electric fields, 69
Sturm, J.C.F., 1
Sturm-Liouville equation, 1
Sturm-Liouville theory, 1
Sturmian basis sets, 11, 110
Sturmian expansion of a plane wave, 13, 111, 112
Sturmian secular equations, 7, 16, 24, 25, 44, 69, 158
Sturmians, xi
Sturmians, Rotenberg's definition, 2
Subgroups, 111
Subspaces, 139, 141
Sum rule, 183
Sum rule for hyperspherical harmonics, 137, 190
Symmetrical form, 140, 170
Symmetry, 16, 33, 38, 44
Symmetry conservation, 194, 195
Symmetry under iteration, 16, 44, 45
Symmetry-adapted basis functions from iteration, 44
Symmetry-adapted basis sets, 16, 26, 27, 43, 44
Symmetry-adapted configurations, 61, 206
Symmetry-adapted secular equations, 27
Synthesis of the wave function, 20
System of N electrons, 13, 193

Target function, 45
Taylor series, 136, 185
Three-center integral, 157
Time-component, 80
Time-independent Dirac equation, 82
Time-independent equation, 19
Time-independent relativistic interelectron interaction, 92
Time-independent solution, 80
Total angular momentum, 26, 43, 84
Total orbital angular momentum, 76
Total solid angle, 125, 137, 183

Total spin, 26, 43
Total spin vector, 76
Transition 4-current, 93
Transition current density, 172
Transition density, 172
Transition dipole moment, 75
Translations, 156
Tree-like graphs, 141
Trees, method of, 139
Trial function, 17, 195
Triatomic molecule, 47
Triplet excited states, 71
Triplet states, 33
Truncated basis set, 16, 44
Turning points, 20
Two-electron operators, 175, 178

Uniform motion, 79
Unit matrix, 31
Unit vector, 129
Unitarity, 27
Uppsala University, xiv

Variational optimization, 26
Vector potential, 79
Velocity of light, 82, 86
Vibrational modes, 47
Virial theorem, 4, 22
Volume element, 112

Weatherford, C., 156
Weighted nuclear charge, 19, 21, 22
Weighted orthonormality, 112
Weighted potential, xi, 5, 11, 161
Weighting factors, xi
Weighting factors, 1, 23, 162
Well ordered eigenvalues, 1
Wulfman, C.E., 154

Zeeman effect, 76
Zero-field dipole moment, 75